Polymers in Solution

Theoretical Considerations
and Newer Methods
of Characterization

Polymers in Solution

Theoretical Considerations
and Newer Methods
of Characterization

Edited by
William C. Forsman
University of Pennsylvania
Philadelphia, Pennsylvania

1986

Plenum Press • New York and London

7302 - 9178

CHEMISTRY

Library of Congress Cataloging in Publication Data

Polymers in solution.

Includes bibliographical references and index.
 1. Polymers and polymerization. 2. Solution (Chemistry) I. Forsman, William C.
QD381.P61245   1986   547.7'0454   86-5090
ISBN 0-306-42146-1

© 1986 Plenum Press, New York
A Division of Plenum Publishing Corporation
233 Spring Street, New York, N.Y. 10013

Printed in the United States of America

Contributors

HERMANN BLOCK ● School of Industrial Science, Cranfield Institute of Technology, Cranfield, Bedford MK43 OAL, United Kingdom.

WILLIAM C. FORSMAN ● Department of Chemical Engineering, University of Pennsylvania, Philadelphia, Pennsylvania 19104.

H. GALINA ● Instytut Technologii Organiccznej, Wroclaw, Poland.

M. GORDON ● The Statistical Laboratory, University of Cambridge, Cambridge CB2 1SB, United Kingdom.

WILLIAM W. GRAESSLEY ● Exxon Research and Engineering Co., Corporate Research—Science Laboratories, Annandale, New Jersey 08801.

JULIA S. HIGGINS ● Department of Chemical Engineering and Chemical Technology, Imperial College, London SW7 2BY, United Kingdom.

L. A. KLEINTJENS ● Research and Patents, DSM, Geleen, Holland.

ANN MACONNACHIE ● Department of Chemical Engineering and Chemical Technology, Imperial College, London SW7 2BY, United Kingdom.

RICHARD A. PETHRICK ● Department of Pure and Applied Chemistry, University of Strathclyde, Glasgow G1 1XL, Scotland, United Kingdom.

B. W. READY ● Department of Chemistry, Wivenhoe Park, Colchester, United Kingdom.

Preface

Polymers in Solution was written for scientists and engineers who have serious research interests in newer methods for characterization of polymer solutions, but who are not seasoned experts in the theoretical and experimental aspects of polymer science. In particular, it is assumed that the reader is not familiar with the development of theoretical notions in conformational statistics and the dynamics of chainlike molecules; how these two seemingly diverse theoretical topics are related; and the role played by polymer–solvent interactions. Chapter 1 thus presents background material that introduces most of the essential concepts, including some of the mathematical apparatus most commonly used in these areas of theory.

This introduction is followed by five chapters that are more closely related to particular experimental techniques. These chapters introduce further theoretical notions as needed. Three of the chapters present considerable detail on the experimental methods, while two other chapters deal more with the interpretation of experimental results in terms of current theories.

Although neutron scattering has become an almost standard technique for the study of conformational properties of macromolecules in the solid state, there has been less emphasis on its application for characterization of polymer molecules in solution. Chapter 4 covers this growing area of application.

The effects of entanglement on the dynamics of polymer chains, and thus on the rheology of their solutions and melts, is an important topic of current interest. Chapter 3 is an account of the state of this area of research. Chapters 2 and 5 deal with another aspect of the dynamics of chain molecules in solution—the shorter wavelength motion of individual repeat units and short blocks of backbone units. Chapter 5 correlates information obtained from NMR, ultrasonic relaxation, and dielectric relaxation, and Chapter 2 treats the new field of dielectric relaxation in a flow field.

Chapter 6 presents an innovative and sensitive method for studying the thermodynamics of polymer solutions: pulse-induced critical scattering (PICS). The method exploits the temperature dependence of light scattered by concentration fluctuations of solutions in the metastable regime to determine the spinodal and cloud point curves.

The book offers no claims to completeness, but the authors feel confident that the material presented here should be useful to those interested in learning more about current work in the study of macromolecules in solution.

W. C. Forsman
Philadelphia

Contents

Statistics and Dynamics of Polymer Molecules in Solution

WILLIAM C. FORSMAN

1. INTRODUCTION

This chapter is intended to serve two purposes. First, it offers a short commentary on theoretical material that can be used as a background for much of the experimental and theoretical discussion given in subsequent chapters. It is aimed at scholars with a serious interest in polymer theory, but not particularly familiar with its development. The coverage is not exhaustive, nor is this chapter intended to be a full-scale technical review. Rather, the intent is to offer a view of the mathematical and physical underpinnings upon which current theoretical and experimental methods are built. Much of the material is thus found in various treatises, the earliest now being almost a classic.[1-3] Little is said here about the newer theoretical developments. These subjects will be introduced at the appropriate places in subsequent chapters, and have been recently reviewed.[4,5]

In most of the topics that are covered, only the essential ideas are presented. In many cases, these seminal ideas were formulated in terms of relatively simple mathematical models and were, at best, semiquantitative in their description of nature. As one might expect, however, the early

WILLIAM C. FORSMAN ● Department of Chemical Engineering, University of Pennsylvania, Philadelphia, Pennsylvania 19104.

models were improved upon without the introduction of radical new ideas. This is, for example, the case in dilute solution thermodynamics and theories of intramolecular and intermolecular excluded volume. Although much has been done—and is still being done—to give these theories more quantitative predictive power, the underlying ideas were clearly formulated as early as 1950. This chapter focuses upon those fundamental ideas.

The second intent of this chapter is to present some of the fundamental notions of conformational statistics and chain dynamics in a way that will bring out certain inherent elements of unity in their mathematical expression. In particular, much of the treatment of conformational statistics given here presents very old ideas that were originally treated by a wide variety of mathematical methods. We demonstrate in this chapter that all of the results from this variety of mathematical methods come out of the mathematics of characteristic functions (Fourier transforms of probability distributions) in a very natural way. Indeed, it is an intent to demonstrate that to know something about characteristic functions is to know a great deal about the conformational statistics of flexible chain molecules.

Expressing conformational statistics in an efficient way by using characteristic functions is a mathematical convenience (and, in this author's experience, an efficient way to organize the material for teaching to students in a graduate course on polymers). The second unifying idea to be discussed in this chapter is, however, far more important because it gets to the heart of the relationship between conformational statistics and chain dynamics. We will show that a set of normal coordinates and their associated eigenvalues exists for any type of chain molecule, linear or branched. These normal coordinates, their eigenvalues, and eigenvectors have special significance for both chain dynamics and conformational statistics. The mean-square radius of gyration of an assembly of the chain molecules, $\langle S^2 \rangle$, can be written as the sum of the mean-squares of their normal coordinates, and the distribution of the radius of gyration, S, is determined by the eigenvalues themselves. The importance of the contribution of each normal coordinate to determining $\langle S^2 \rangle$ and the distribution of S drops off rapidly with its order—indeed, to a first approximation, the conformational statistics are determined by only the first few normal coordinates and their eigenvalues. The equation for the dynamics of free-draining chain molecules are separated by a transformation to these same normal coordinates, and there is a one-to-one correspondence between the relaxation times given by the Rouse model of chain molecules and the contribution to chain motion from each normal coordinate.

The zero-shear-rate viscosity of solutions of free-draining chain molecules is the sum of contributions from each of the normal coordinates. Just as in the case of conformational statistics, the contribution of each normal coordinate to the viscosity drops off rapidly with its order and, to a first approximation, the viscosity is determined only by the first few normal coordinates.

Finally, we present a simple prescription derived from graph theoretical methods that gives the normal coordinates and their eigenvalues for any form of chain branching. It will be shown that, since various types of branched chains could have the same set of eigenvalues, we must conclude that experiments that give information on chain dynamics and conformational statistics give information only on the eigenvalues and not on the detailed nature of the branching per se.

2. MATHEMATICAL PRELIMINARIES

Most readers have been exposed to many of the mathematical methods used in this book—both in this chapter and in the following chapters more closely associated with experimental techniques. Nonetheless, there are two methods that play such an important role that it might be worthwhile giving them special attention. Much of the development in this chapter is presented in matrix algebra form. Since this topic is now standard fare, even in undergraduate programs, only a brief summary is given. Indeed, only some of the basic notions are required for the level of presentation given in this chapter: a comfortable understanding of matrix multiplication, the definition of the inverse, and the procedures of orthogonal transformation and diagonalization of symmetric matrices.

Section 7 of this chapter is entitled Graph Theory and the Statistics and Dynamics of Random-Flight Chains. The title may imply that the material might be "heavy going," but it is not. Only a few definitions from graph theory are used. All of the rest is, in fact, nothing more than matrix algebra. This is not to say that sophisticated graph theoretical methods have not been used in this class of theoretical problems. Although reference will be made to this work, we will not go into it in detail.

The second special topic, Fourier transforms and their application to probability theory, is not quite so commonly found in the store of mathematical tools used by scientists and engineers working in polymers. Because of the extraordinary power of this method and its almost all-pervasive use in polymer theory, it is worth going over some of the fundamental notions and relationships in some detail.

There is an abundance of reference material covering these topics. The author's favorites, and those from which the material in this section was drawn, are cited at the beginning of each section.

2.1. Matrix Algebra[6,7]

Matrices will be indicated in boldface type such as $\mathbf{A}$. If a matrix $\mathbf{A}$ is an array of p rows and q columns, it will be referred to as a p by q matrix. Its elements will be written as a_{ij}, where i indicates the row and j the column of the position of the element in the array. The transpose of matrix $\mathbf{A}$ will be denoted by $\mathbf{A}^T$ and its elements by a_{ij}^T. From the definition of the transpose, $a_{ij}^T = a_{ji}$. Column and row matrices are the exception to the two-subscript convention. Unless there is an indication to the contrary, collections of variables will be represented as a column matrix—often referred to as a column vector. It is important to keep in mind that we will be dealing with column vectors with many elements as well as the usual three-dimensional (directed) line vectors. A column matrix (or vector) and its transpose (a row matrix or vector) will be written as follows:

$$
\mathbf{x} = \begin{bmatrix} x_1 \\ x_2 \\ x_3 \\ \cdot \\ \cdot \\ x_n \end{bmatrix}, \qquad \mathbf{x}^T = [x_1 \ x_2 \ x_3 \ \ldots \ x_n] \qquad (2.1)
$$

Matrix addition and subtraction is written $\mathbf{C} = \mathbf{A} \pm \mathbf{B}$, where $c_{ij} = a_{ij} \pm b_{ij}$. Multiplication of $\mathbf{A}$ times $\mathbf{B}$ to give matrix $\mathbf{C}$ is written $\mathbf{AB} = \mathbf{C}$, where

$$
c_{ij} = \sum_{k=1}^{q} a_{ik} b_{kj} \qquad (2.2)
$$

An implication in eq. (2.2) is that two matrices must be conformable for the operation of multiplication to be defined. The matrices need not be square, but if $\mathbf{A}$ is a p by q matrix, $\mathbf{B}$ must be a q by r matrix and the product matrix will be p by r. In general, the result of matrix multiplication depends upon the order the two matrices are written, i.e., multiplication is, in general, not commutative.

Multiplication of a matrix $\mathbf{B}$ by a constant a to give a new matrix $\mathbf{C}$ is denoted by $\mathbf{C} = a\mathbf{B} = \mathbf{B}a$, and the elements of $\mathbf{C}$ are given by

$$c_{ij} = ab_{ij} \tag{2.3}$$

An important relationship involving the transpose of the product of matrices is

$$(\mathbf{ABC})^T = \mathbf{C}^T\mathbf{B}^T\mathbf{A}^T \tag{2.4}$$

Matrix multiplication is a convenient device for the generation of quadratic forms. The following operation is typical:

$$\begin{aligned}
\mathbf{X}^T\mathbf{AX} = {} & a_{11}x_1{}^2 + a_{12}x_1x_2 + a_{13}x_1x_3 + \cdots \\
& + a_{21}x_2x_1 + a_{22}x_2{}^2 + a_{23}x_2x_3 + \cdots \\
& + a_{31}x_3x_1 + a_{32}x_3x_2 + a_{33}x_3{}^2 + \cdots
\end{aligned} \tag{2.5}$$

A quadratic form is a scalar and not a matrix (unless one would like to think of it as a matrix of one element). In the above example the expression on the right-hand side is written in "matrixlike" form only to indicate how the terms are generated from the matrix multiplication shown on the left-hand side; clearly the x_1x_j and x_jx_i terms would be combined in a "normal" writing of the expression.

Determinants can be defined for square matrices. A square matrix for which the value of the determinant is not zero is said to be nonsingular. Every nonsingular matrix $\mathbf{A}$ has a unique inverse, which is written $\mathbf{A}^{-1}$, where

$$\mathbf{AA}^{-1} = \mathbf{A}^{-1}\mathbf{A} = \boldsymbol{\delta} \tag{2.6}$$

The elements of $\boldsymbol{\delta}$ are the Kronecker deltas, that is,

$$\begin{aligned}
\delta_{ij} &= 1 \qquad \text{if } i = j \\
&= 0 \qquad \text{if } i \neq j
\end{aligned} \tag{2.7}$$

It is convenient to effect a transformation from one set of variables represented by a column matrix $\mathbf{x}$ to another set $\mathbf{x}'$ by the operation

$$\mathbf{x}' = \mathbf{Dx}, \qquad \mathbf{x} = \mathbf{D}^{-1}\mathbf{x}' \tag{2.8}$$

The operation is said to be an orthogonal transformation if $\mathbf{D}^T = \mathbf{D}^{-1}$, and $\mathbf{D}$ is said to be an orthogonal matrix. It is also true that any symmetric matrix $\mathbf{A}$ can be transformed into a diagonal matrix (that is, one with nonzero elements only on the diagonal) by the following operation using an appropriate orthogonal matrix **D**:

$$\mathbf{D}^{-1}\mathbf{A}\mathbf{D} = \mathbf{\Omega} \tag{2.9}$$

The diagonal elements of $\mathbf{\Omega}$ are abbreviated $\Omega_{jj} \equiv \Omega_j$, and are called the eigenvalues of **A**. It is important to note that there is only *one* orthogonal transformation that will effect the diagonalization of a symmetric matrix.

2.2. Fourier Transforms[8,9]

The Fourier transform of a function $f(x)$ is written $\bar{f}(\omega)$ and is defined by the integral

$$\bar{f}(\omega) = \int_{-\infty}^{+\infty} f(x)e^{-i\omega x}\,dx \tag{2.10}$$

The Fourier transform of $f(x)$ exists only if the above integral exists for every real value of ω. If the Fourier transform $\bar{f}(\omega)$ exists, it is always possible to express the original function by the following integral relationship:

$$f(x) = \frac{1}{2\pi} \int_{-\infty}^{\infty} \bar{f}(\omega)e^{i\omega x}\,d\omega \tag{2.11}$$

Evaluation of this integral to recover the function $f(x)$ is known as inverting the transform.

Deriving Fourier transforms and inverting them is an interesting mathematical exercise, but that is not why a discussion of this process is included in this book. Transform mathematics is intimately mixed with theory and experiment in the physical and engineering sciences. In many theoretical developments, the sought-after relationship does not come out of the mathematical manipulations directly, but emerges as a Fourier transform. The transform must be inverted before the problem is solved. In some experimental procedures such as diffraction of x-rays and neutrons and determining linear viscoelastic behavior, the experimentally measured quantities give the Fourier transform of the physical property that is being explored and not the property itself. Again, the problem is not solved until the transform has been inverted.

Considering the importance of Fourier transforms in the physical sciences and engineering, we should be careful in matters of their existence. This is a rather involved subject, and we will not take it up here. It suffices to say, however, that virtually all functions one is likely to encounter in the physics and chemistry of polymers have transforms. They can have discontinuities (as long as there is a finite number of them). They can have infinities (as long as they are well enough behaved). Constants have transforms, and polynomials of all orders have transforms. There are, however, important ground rules in evaluating the integrals in Eqs. (2.10) and (2.11) if all of this is to be true.

The first ground rule has to do with the infinite limits in the integrals. We must remember that integrals of this form are known as *improper* integrals. An improper integral represents the limit of a "proper" integral as its range of integration goes to infinity. That is,

$$\int_{-\infty}^{\infty} F(x)\,dx = \lim_{\substack{a \to \infty \\ b \to \infty}} \int_{-a}^{b} F(x)\,dx \tag{2.12}$$

The implication in equation (2.12) is that the integral is first evaluated over the range from $-a$ to b, where a and b have fixed, positive values. Then, if the resulting integrals converge as a and b approach infinity *independently*, the improper integral is said to exist. This criterion for the existence of an improper integral is, however, too general for Fourier transform mathematics. In order to include the broadest spectrum of functions for which Fourier transforms exist, we must evaluate the integrals in Eqs. (2.10) and (2.11) according to the rule which gives their Cauchy principal value. Cauchy's principal value of an improper integral is defined as follows:

$$\int_{-\infty}^{\infty} F(x)\,dx = \lim_{b \to \infty} \int_{-b}^{b} F(x)\,dx \tag{2.13}$$

A second form of improper integral is often encountered in practice. In this case, the singularity in the integrand is at the point $x = a$. In particular, we are interested in singularities where $f(x) \to \infty$ as $x \to a$, where a is a finite constant. A general definition for the existence of such an improper integral is the existence of the limit

$$\lim_{\substack{\varepsilon \to 0 \\ \delta \to 0}} \left[\int_{-\infty}^{a-\varepsilon} f(x)\,dx + \int_{a+\delta}^{\infty} f(x)\,dx \right] \tag{2.14}$$

where ε and δ approach zero independently. The above limit may not exist. But, whether or not it does, the Cauchy principal value of such an integral is defined by

$$\lim_{\to 0} \left[\int_{-\infty}^{a-\varepsilon} f(x)\,dx + \int_{a+\varepsilon}^{+\infty} f(x)\,dx \right] \tag{2.15}$$

Finally, we should add that both types of improper behavior (as $x \to \infty$ and as $x \to a$) may appear in a single improper integral, and limits both to ∞ and to a must be taken. An example of such behavior will be shown shortly.

The difference between these two types of definitions of the value of improper integrals may seem small, but it is often the difference between the existence and nonexistence of Fourier transforms for many functions.

The second ground rule in evaluating and inverting transforms has to do with the order in which parameters (such as the limits of improper integrals) are allowed to go to infinity when we are dealing with more than one. This point will not be pursued in detail here, but it is important to note, however, that care must be taken or the operation will give total nonsense. This will be illustrated in examples to follow.

2.2.1. A Simple Example

For the first example we will pick $f(x)$ to be equal to the sign of x, that is,

$$f(x) = \text{sgn}(x) = x/|x| \qquad \text{if } x \neq 0 \tag{2.16}$$

Since the function is equal to -1 if $x < 0$ and $+1$ if $x > 0$, there is a finite discontinuity at $x = 0$. The transform of this function is, however, easy to evaluate by integrating over the two parts of $f(x)$ as follows:

$$\tilde{f}(\omega) = \int_{-\infty}^{0} (-1)e^{-i\omega x}\, dx + \int_{0}^{\infty} (+1)e^{-i\omega x}\, dx$$

$$= \frac{-2i}{\omega} + 2i\{\lim_{M \to \infty} [\cos(\omega M)/\omega]\} \tag{2.17}$$

The first contribution to $\tilde{f}(\omega)$ is an ordinary function, whereas the second contribution is pathological in that it becomes an infinitely rapidly oscillating function as $M \to \infty$. Nonetheless, given the transform of $\text{sgn}(x)$, including the pathological part, it is instructive to test the inversion relationship to see if it can, in fact, generate the function, complete with the discontinuity at $x = 0$.

We will look at the inversion of the second part of the transform first. Indeed, we will examine the behavior of the pathological part of the function when it is used in the only way it is defined—as a factor of an integrand. We will thus test it with any sufficiently well-behaved function $f(\omega)$. If the expression were to represent the inversion of the transform in question, the function $\bar{f}(\omega)$ would be $\exp(i\omega x)$. For the moment we will ignore the factor $2i$. Since $[\cos(\omega M)/\omega]$ is singular at $\omega = 0$, we must write an integral covering the entire range of ω in the following way:

$$\lim_{\substack{M\to\infty \\ b\to\infty \\ \varepsilon\to 0}} \left[\int_{-b}^{-\varepsilon} f(\omega) \frac{\cos(\omega M)}{\omega} \, d\omega + \int_{+\varepsilon}^{+b} f(\omega) \frac{\cos(\omega M)}{\omega} \, d\omega \right] \qquad (2.18)$$

If we introduce the change in variable $y = \omega M$, the above integral becomes

$$\lim_{\substack{M\to\infty \\ b\to\infty \\ \varepsilon\to 0}} \left[\int_{-bM}^{-\varepsilon M} f(y/M) \frac{\cos(y)}{y} \, dy + \int_{+\varepsilon M}^{+\varepsilon M} f(y/M) \frac{\cos(y)}{y} \, dy \right] \qquad (2.19)$$

We will replace the parameters in the limits of the integrals by the new set:

$$\varepsilon' = \varepsilon M$$
$$b' = bM \qquad\qquad (2.20)$$

and allow M to increase without limit while keeping ε' and b' finite. For this operation the function in the integrand goes to $f(0)$, which is a constant that can be removed from under the integrand and factored from the expression. The result is

$$\lim_{\substack{b'\to\infty \\ \varepsilon'\to 0}} f(0) \left[\int_{-b'}^{-\varepsilon'} \frac{\cos(y)}{y} \, dy + \int_{+\varepsilon'}^{b'} \frac{\cos(y)}{y} \, dy \right] \qquad (2.21)$$

Since $[\cos(y)/y]$ is an odd function, the two terms in the above square brackets cancel each other out, and the result is equal to zero regardless of the values of b' and ε'. We thus see that the second part of the transform as written in Eq. (2.17) contributes nothing to the inversion. Within the framework of transform mathematics, then, we can set the pathological contribution equal to zero, that is,

$$\lim_{M\to\infty} [\cos(\omega M)/\omega] \equiv 0 \qquad (2.22)$$

We can thus turn our attention to determining the contribution to the first part of the transform from Eq. (2.17). Since the term $(-2i/\omega)$ is the only nonzero part of the result when we invert the transform, we can write the inversion simply as

$$f(x) = \frac{1}{2\pi} \int_{-\infty}^{\infty} \frac{-2i}{\omega} e^{i\omega x} \, d\omega \qquad (2.23)$$

Replacing the complex exponential with its trigonometric equivalent and remembering that we must evaluate the Cauchy principal value of the integral gives

$$f(x) = \lim_{b \to \infty} \frac{1}{2\pi} \int_{-b}^{b} \frac{-2i}{\omega} \left[\cos(\omega x) + i \sin(\omega x) \right] d\omega \qquad (2.24)$$

We note that $\cos(\omega x)/\omega$ is an odd function of ω. When we take the integral from $-b$ to $+b$ the contribution to the integral from the cosine term thus vanishes even before we take the limit as $b \to \infty$. Indeed, we notice that if the result of the integration is to be a real number, the contribution from the $\cos(\omega x)$ term must vanish upon inversion of the transform. The other contribution to the integrand, $\sin(\omega x)/\omega$, is an even function and does indeed contribute to the result:

$$f(x) = \lim_{b \to \infty} \frac{1}{\pi} \int_{-b}^{+b} \frac{\sin(\omega x)}{\omega} d\omega \qquad (2.25)$$

The integration is easily done by introducing the change in variable

$$y = \omega x \qquad (2.26)$$

which gives

$$f(x) = \lim_{b \to \infty} \frac{1}{\pi} \int_{-bx}^{bx} \frac{\sin(y)}{y} dy \qquad (2.27)$$

It is important to remember that we must fix the value of x first and then take the limit as $b \to \infty$. It is clear that the result will depend upon the sign of x. If $x > 0$, letting b go to infinity gives limits of the integration from $-\infty$ to $+\infty$, and it is well known that

$$\int_{-\infty}^{\infty} \frac{\sin(y)}{y} dy = \pi \qquad (2.28)$$

For positive values of x, then, we find that $f(x) = 1$.

On the other hand, if $x < 0$ the situation is quite different. As b goes to infinity, the limits of integration go from $+\infty$ to $-\infty$, and the sign of the integral is reversed. The value of the integral is then $-\pi$, and $f(x) = -1$. The case where $x = 0$ is different still. For that value of x, the limits of integration are "from 0 to 0"—that is, the range of integration vanishes and the value of the integral is zero. We find, then, that inversion gives us the following definition of $f(x)$:

$$\begin{aligned} f(x) &= 1 & \text{for } x > 0 \\ &= -1 & \text{for } x < 0 \\ &= 0 & \text{for } x = 0 \end{aligned} \qquad (2.29)$$

One should notice that something was missing in the original definition of $f(x)$. We did not stipulate the value of the function at the point of the discontinuity, $x = 0$. In point of fact, we could have defined $f(0)$ as any value we had wanted, but by the definition of the Riemann integral for a function with such a discontinuity, it would have not effected the value of the integral. So no matter how we might have defined $f(0)$, we would have had the same transform for inversion, and would have gotten the same value for $f(0)$. Indeed, this is a general property of Fourier transforms—when the transform of a function with a finite discontinuity is inverted, the inversion gives a value for the function at the discontinuity which is the average of the limits of the function as it approaches the discontinuity from either side.

2.2.2. A Slightly More Involved Example

For the second example, we will assume that $f(x)$ is equal to $1/2l$ between $x = -l$ and $x = +l$ and zero everywhere else. Application of Eq. (2.10) to this example gives

$$\tilde{f}(\omega) = \frac{1}{2l} \int_{-l}^{+l} e^{-i\omega x} \, dx = \frac{\sin(\omega l)}{\omega l} \tag{2.30}$$

Obtaining the transform of this function was straightforward. The more interesting question, however, is how the function can be recovered along with its two discontinuities when Eq. (2.30) is substituted into Eq. (2.11). If we proceed we get

$$f(x) = \frac{1}{2\pi} \int_{-\infty}^{\infty} \frac{\sin(\omega l)}{\omega l} e^{i\omega x} \, d\omega \tag{2.31}$$

As in the first example, it is convenient to write the complex exponential in the equivalent trigonometric form, which gives

$$f(x) = \frac{1}{2\pi} \int_{-\infty}^{\infty} \frac{\sin(\omega l)}{\omega l} [\cos(\omega x) + i \sin(\omega x)] \, d\omega \tag{2.32}$$

Again, it is important to inspect the integrand for even and odd functions. In this case, $\sin(\omega l)/\omega l$ and $\cos(\omega x)$ are even functions of ω and $\sin(\omega x)$ is an odd function of ω. Since an even function times an even function is an even function, and an even function times an odd function is an odd function, the contribution to the integral from the second term in the square brackets vanishes (which, of course it must if the inversion is to give a real function). All that remains, then, is

$$f(x) = \frac{1}{2\pi} \int_{-\infty}^{\infty} \frac{\sin(\omega l)\cos(\omega x)}{\omega l} \, d\omega \qquad (2.33)$$

A trigonometric identity can be used to convert the product of sine and cosine terms in the numerator into a more useful form for integration. At the same time, we will acknowledge that it is the Cauchy principal value that we must calculate by rewriting the above expression in the following form:

$$f(x) = \lim_{M \to \infty} \frac{1}{2\pi} \int_{-M}^{M} \frac{\frac{1}{2}[\sin(\omega l + \omega x) + \sin(\omega l - \omega x)]}{\omega l} \, d\omega \qquad (2.34)$$

The above integral is made up of contributions from the two sine terms. If we introduce the changes of variable

$$y = \omega(l \pm x) \qquad (2.35)$$

(with the plus or minus sign being used depending upon which sine function is being considered), we get a pair of integrals of the form

$$\lim_{M \to \infty} \frac{1}{4\pi} \int_{-M(l \pm x)}^{M(l \pm x)} \frac{\sin(y)}{ly} \, dy \qquad (2.36)$$

Since the value of x is fixed before the parameter M goes to infinity, the results of each of the integrations depend upon the value of x. If $x < -1$, the sum of the two contributions is

$$\lim_{M \to \infty} \frac{1}{4\pi} \int_{M}^{-M} \frac{\sin(y)}{ly} \, dy + \lim_{M \to \infty} \frac{1}{4\pi} \int_{-M}^{M} \frac{\sin(y)}{ly} \, dy = 0 \qquad (2.37)$$

The same result is obtained if $x > +l$. If, on the other hand, $-l < x < +l$, each of the two contributing integrals becomes

$$\lim_{b \to \infty} \frac{1}{4\pi} \int_{-b}^{b} \frac{\sin(y)}{ly} \, dy = \frac{1}{4l} \qquad (2.38)$$

Since there are two contributions, the value of the function in this interval is given by

$$f(x) = 1/2l \qquad (2.39)$$

just as it should be.

We must finally consider the cases where $x = \pm l$. For $x = -l$ both the lower and upper limits in the first contribution to the integral in Eq. (2.36) are exactly equal to zero, and that contribution vanishes. The lower and upper limits in the second contribution become $-\infty$ and $+\infty$, and the result is thus

$$f(x) = 1/4l \qquad (2.40)$$

The same result is obtained for $x = +l$, except in this case it is the second contribution which vanishes. Again we see that when we invert the Fourier transform of a function with discontinuities, the inversion gives values for the function at the discontinuities that are just the averages of the limits that the function approaches from the two different directions.

2.2.3. The Dirac Delta Function

Combining the concepts of the Dirac delta function and the Fourier transform gives a powerful set of mathematical tools. The Dirac delta function is often considered to be the mathematical idealization of an impulse. The implication of this concept is that the delta function is very large at some value of the independent variable x, and zero everywhere else. When the impulse is to be applied at the origin ($x = 0$), the delta function is denoted by $\delta(x)$, and it is sometimes said that $\delta(x) = \infty$ if $x = 0$ and $\delta(x) = 0$ for all other values of x. Although this picture of the delta function has some intuitive appeal, it is not a rigorous mathematical definition. The delta function is, in fact, not a function at all but an operation performed using any one of many functions in the limit as an appropriate parameter associated with the function approaches some limit, usually zero or infinity. Consider the following operations and definitions.

We will define what we will call a precursor to the delta function, $\delta_a(x)$. The parameter, here denoted as a, will define the exact details of the precursor function, but a could appear in different ways in the various precursors. Writing the delta function in this form implies that the function (or impulse) is centered at the position $x = 0$. To make the treatment more general, we will write the delta function $\delta_a(x - x_0)$, which implies that the function is shifted a distance x_0 to the right, or centered over the position $x = x_0$. (If x_0 is negative, the precursor function has been shifted to the left.)

For the moment we will not be overly concerned about the exact form of the precursor function, but it will be assumed that it includes the arbitrary parameter a. An important aspect in the definition of a prototype function is, however, that it must be defined in such a way that

$$\int_{-\infty}^{\infty} \delta_a(x - x_0)\,dx = 1 \qquad (2.41)$$

regardless of the value of a. If $\delta_a(x)$ is sufficiently well behaved, the above relationship implies that the area under the curve is equal to unity, no matter what the value of a. In addition, the prototype function must be defined in such a way that

$$\lim_{a \to \infty} \int_{-\infty}^{\infty} f(x)\delta_a(x - x_0)\,dx = f(x_0) \qquad (2.42)$$

In actual practice, however, the following "shorthand" notation is used to denote the above operation:

$$\int_{-\infty}^{\infty} f(x)\delta(x - x_0)\,dx = f(x_0) \qquad (2.43)$$

It should be noted, that to be consistent with the convention used in Fourier transforms, the integral in Eq. (2.43) must be the Cauchy principal value.

One such precursor function is

$$\delta_a(x - x_0) = (1/2a) \qquad \text{for } x_0 - a \leq x \leq x_0 + a$$
$$= 0 \qquad \text{everywhere else} \qquad (2.44)$$

When used to generate a delta function, the parameter a is allowed to approach zero. This simple example demonstrates the pathological nature of delta functions generated from their precursors. For the moment we will ignore the fact that $\delta_a(x - x_0)$ is intended to be used only in conjunction with an integral as in Eq. (2.42). If we proceed to take the limit as $a \to 0$, we see that $\delta_a(x - x_0)$ does, in fact, become infinite at the point $x = x_0$ and zero everywhere else—hardly the sort of function one would call well behaved. If, however, the integration is performed *first*, and *then* the limit is taken, this form of $\delta_a(x - x_0)$ leads to the well-defined result described in Eq. (2.43).

Another precursor for the delta function is

$$\delta_\beta(x - x_0) = (\beta/\sqrt{\pi})e^{-\beta^2(x-x_0)^2} \qquad (2.45)$$

where the parameter, this time denoted as β, is taken to infinity in Eq. (2.42).

2.2.4. The Fourier Transform of the Delta Function

By applying the prescriptions set down in Eqs. (2.10) and (2.42), one can determine the Fourier transform of the delta function. The result is

$$\bar{\delta}(\omega) = \int_{-\infty}^{\infty} \delta(x - x_0)e^{-i\omega x} \, dx = e^{-i\omega x_0} \qquad (2.46)$$

If the delta function is located at the origin, i.e., $x_0 = 0$, its Fourier transform is just equal to unity.

Since the Fourier transform of the delta function is $e^{-i\omega x_0}$, we should be able to substitute this expression into Eq. (2.11) and recover the delta function itself. When we make this substitution we get

$$\delta(x - x_0) = (1/2\pi) \int_{-\infty}^{\infty} e^{i\omega[x-x_0]} \, d\omega \qquad (2.47)$$

It is not clear that this improper integral is indeed a legitimate form of the delta function. The situation begins to clear up, however, if we follow the proper prescription and take the Cauchy principal value of the integral, and write

$$\delta(x - x_0) = \lim_{R \to \infty} (1/2\pi) \int_{-R}^{R} e^{i\omega[x-x_0]} \, d\omega \qquad (2.48)$$

We thus see that the parameter, R, required for the evaluation of the Cauchy principal value becomes the parameter in the precursor of the delta function. It is thus appropriate in this case to write the integral form of the precursor as $\delta_R(x - x_0)$.

The precursor of the delta function defined in Eq. (2.48) bears little resemblance to those given in Eqs. (2.44) and (2.45). More insight into the nature of this expression can, however, be obtained by evaluating the integral. It can, in fact, be written in simple closed form as

$$\delta_R(x - x_0) = \frac{\sin[R(x - x_0)]}{\pi(x - x_0)} \qquad (2.49)$$

If we multiply both the numerator and denominator by R we get

$$\delta_R(x - x_0) = (R/\pi) \frac{\sin[R(x - x_0)]}{R(x - x_0)} \qquad (2.50)$$

In testing whether or not the expression in Eq. (2.49) is a proper precursor for a delta function, we see if it satisfies Eq. (2.41). The integration is straightforward, and it does. Other than that, however, it does not seem to be properly behaved. If $x = x_0$, the value of $\sin[R(x - x_0)]/[R(x - x_0)]$ is equal to unity regardless of the value of R. We thus see that as $R \to \infty$, the precursor goes to infinity as (R/π) at $x = x_0$—just as we would expect. On the other hand, if $x \neq x_0$ the precursor does *not* go to zero with increasing R as the other precursors we considered, but only oscillates with increasing frequency. Intuitively, however, one can see how this oscillation at infinite frequency gives the behavior stipulated in Eq. (2.42). With any sufficiently well-behaved test function, the extremely rapidly oscillating $\sin[R(x - x_0)]$ would go through many maxima and minima with negligible change in the test function or the denominator, $\pi(x - x_0)$. The contribution to the integral in Eq. (2.42) would sum to zero except in the neighborhood of the point $x = x_0$. This intuitive perception is not, however, sufficiently convincing for one to feel comfortable with the notion that the expression in Eq. (2.49) could actually serve as the precursor for a Dirac delta function. That it is, however, is easily demonstrated by applying the basic definitions. We begin by examining the integral:

$$\lim_{R \to \infty} \frac{1}{\pi} \int_{-\infty}^{\infty} f(x) \left[\frac{\sin[R(x - x_0)]}{(x - x_0)} \right] dx \tag{2.51}$$

If we make the change in variable, $y = R(x - x_0)$, take the Cauchy principal value of the above integral, and remember that the precursor becomes the delta function as $R \to \infty$, we get

$$\lim_{\substack{R \to \infty \\ N \to \infty}} \frac{1}{\pi} \int_{-N}^{N} f\left(\frac{y}{R} + x_0 \right) \frac{\sin(y)}{y} dy \tag{2.52}$$

If we let R go to infinity first (i.e., with y held constant), the test function goes to $f(x_0)$, which is independent of y and can be removed out from under the integral sign. Then, when we let N go to infinity the value of the integral is π. The value of integral (2.51) is thus $f(x_0)$, and it has been demonstrated that the precursor from Eq. (2.49) does indeed generate a delta function.

2.2.5. Convolution

Consider two functions $f_1(x)$ and $f_2(x)$, and a third function $f_{2c}(x)$ defined by the integral

$$f_{2c}(x) = \int_{-\infty}^{\infty} f_1(y) f_2(x - y) \, dy \tag{2.53}$$

The function f_{2c} is said to be the *convolution* of functions f_1 and f_2. It is easy to show that the results of a convolution are independent of the order in which we take the functions under the integral sign. That is, $f_{2c}(x)$ is equally well given by

$$f_{2c}(x) = \int_{-\infty}^{\infty} f_2(y) f_1(x - y)\, dy \qquad (2.54)$$

The subscript $2c$ implies that f_{2c} is the convolution of *two* functions, and the following notation is sometimes used to stress that the results of the integration are independent of which function is taken as being a function of y and which as a function of x–y:

$$f_{2c}(x) = f_1(x) * f_2(x) \qquad (2.55)$$

Once $f_{2c}(x)$ has been defined, it is possible to define a new function $f_{3c}(x)$ by the convolution of $f_{2c}(x)$ and a third function $f_3(x)$. As one would suspect, the order with which any of the functions are convoluted with respect to either of the other two is independent of the order. It is therefore proper to write

$$f_{3c}(x) = f_1(x) * f_2(x) * f_3(x) \qquad (2.56)$$

Indeed the convolution of any number of functions can be formed with a result that is independent of order. We can thus write

$$f_{nc}(x) = f_1(x) * f_2(x) * \cdots * f_n(x) \qquad (2.57)$$

Taking the Fourier transform of $f_{2c}(x)$ gives an interesting result:

$$\tilde{f}_{2c}(\omega) = \int_{-\infty}^{\infty} f_1(y) \int_{-\infty}^{\infty} f_2(x - y) e^{-i\omega x}\, dy\, dx \qquad (2.58)$$

If we effect the change in variable $z = x - y$, Eq. (2.58) can be written

$$\tilde{f}_{2c}(\omega) = \int_{-\infty}^{\infty} f_1(y) e^{-i\omega y}\, dy \int_{-\infty}^{\infty} f_2(z) e^{-i\omega z}\, dz \qquad (2.59)$$

Not only does the change in variable reduce the double integral to the product of two integrals, we recognize them as the Fourier transforms of $f_1(x)$ and $f_2(x)$, respectively. We thus find that

$$\tilde{f}_{2c}(\omega) = \tilde{f}_1(\omega) \tilde{f}_2(\omega) \qquad (2.60)$$

It is, of course, irrelevant in which order the product is taken.

This concept is easily extended to any number of functions; it can be shown that if a function is generated by a multiple convolution of other functions, its Fourier transform is given by

$$\bar{f}_{nc}(\omega) = \bar{f}_1(\omega)\bar{f}_2(\omega)\bar{f}_3(\omega) \cdots \bar{f}_n(\omega) \tag{2.61}$$

2.3. Probability Theory[10,11]

In many areas of science and engineering the act of measuring some physical quantity will, in principle, give the experimenter some precise numerical value, even if that value can be determined only to within some experimental error. Unfortunately, this is not often the case in polymer science. Consider the following hypothetical examples. If one could pick a molecule out of a flask at the completion of a polymerization reaction and measure its molecular weight, that measurement would tell little about the molecular weight of the rest of the molecules. Likewise, if we could determine the end-to-end distance of one polymer molecule out of a collection of molecules of the same molecular weight, the result would tell us little about the end-to-end distances of the rest. If, however, we examined many molecules out of the reaction flask, we could infer much about the molecular weights produced by the reaction. Likewise, if we could measure the end-to-end distance of many molecules out of the collection of molecules of the same molecular weight, we would clearly be in a position to say something quantitative about end-to-end distances.

This added complexity in dealing with the physical world is handled by interpreting hypothetical acts such as picking a molecule out of a reaction flask and measuring its molecular weight, or measuring the end-to-end distance of one molecule out of a collection of molecules as sampling a *random variable*. For our purposes, we will consider a random variable as a number obtained as the outcome of making a well-defined observation. In probability theory, the act of making such an observation is often referred to as making a *trial*. The concept of making an observation is a general one, and is not impeded by the fact that it might be very difficult to make measurements such as the ones mentioned above. Random variables most commonly discussed in treatises on probability theory are the outcome of flipping a coin or rolling dice. In these examples, one must define the numerical value associated with the outcome. For example, in flipping a single coin we could define the random variable as X. We could then say that if a flip of the coin comes up heads, the measurement gives the result $X = +1$, and if it comes up tails, the observation gives $X = -1$. One might elect to call the outcome of rolling one die Y, and further elect to define the value of Y as a number from one to six, depending upon which face comes out on top after a roll.

The outcomes from coin flipping and die rolling are examples of discrete random variables—that is, a trial can give only one outcome out of a discrete set of possible outcomes. Likewise, picking a molecule out of a reaction flask is sampling a discrete random variable. The outcome of a trial can give only a monomer, a dimer, or a molecule of some higher molecular weight. In this example, the numerical value of the outcome is, of course, the molecular weight of the selected molecule. The end-to-end distance of a linear macromolecule is an example of a continuous random variable. In principle, one could find molecules with any value between zero and the length of the fully extended chain.

2.3.1. Probability Distributions

Both discrete and continuous random variables have associated probability distributions. For a discrete random variable the distribution is a set of discrete probabilities which are often written $p(X = X_i)$. The operational interpretation of this concept is as follows. Assume that one makes N trials, out of which there are N_i outcomes of each possibility i. Then, if we let the number of trials increase without limit, by definition

$$\lim_{N \to \infty} \frac{N_i}{N} = p(X = X_i) \qquad (2.62)$$

It is important to note, however, that after a trial is made the population is returned to its original state—that is, if a molecule is removed from a reaction flask for molecular weight determination it is returned after the measurement has been made. Alternatively, we can assume that the limit in Eq. (2.62) is approached sufficiently closely long before removing molecules has made any impact upon the nature of the system.

In general, the notation $p(\ )$ will be interpreted as the probability of observing the outcome described within the parentheses. In the case of discrete random variables, it is, however, convenient to introduce the following simplified notation:

$$P_i \equiv p(X = X_i) \qquad (2.63)$$

The set of numbers P_i will be defined as the probability distribution for the discrete random variable. It follows from the above definitions that

$$\sum P_i = 1 \tag{2.64}$$

Probability distributions are defined somewhat differently for continuous random variables. In this case a distribution function $P(X)$ is defined such that

$$p(X_1 < X < X_2) = \int_{X_1}^{X_2} P(X) \, dX \tag{2.65}$$

From the operational standpoint, $p(X_1 < X < X_2)$ represents the fraction of observations of the random variable that are found between X_1 and X_2 as the number of trials approaches infinity. From these definitions it follows that

$$\int_{-\infty}^{\infty} P(X) \, dX = 1 \tag{2.66}$$

As in the case of discrete random variables, we must be careful about the physical interpretation of allowing the number of trials to increase indefinitely.

2.3.2. Moments, Means, and Expectation Values

The nth moment of a discrete distribution, $\langle X^n \rangle$, is defined by the relation

$$\langle X^n \rangle = \sum X_i^n P_i \tag{2.67}$$

where the summation is taken over all of the possible discrete outcomes. For a continuous random variable, Eq. (2.67) is replaced by

$$\langle X^n \rangle = \int_{-\infty}^{\infty} X^n P(X) \, dX \tag{2.68}$$

These quantities are equal to mean values determined by sampling the population of a discrete or continuous random variable as the number of trials is increased without limit. Since, strictly speaking, one can never effect an infinite number of trials, these quantities are sometimes referred to as the expectation values of the quantities X^n. In point of fact, when sampling the population represents examining about 10^{20} molecules, there is no distinction between the nth moment and the expectation value of X^n.

2.3.3. Discrete Distributions Revisited

It is useful to reconsider the definition of probabilities for discrete random variables. The Dirac delta function can be used to define distribution functions for discrete cases by writing

$$P(X) = \sum P_i \delta(X - X_i) \tag{2.69}$$

Incorporation of the above definition into probability theory allows us to use the formal language of continuous random variables for both the continuous and discrete cases. Indeed, it is clear that all of the normalization and moment relationships are obeyed when we use this convention.

2.3.4. Characteristic Functions and Moment Expansions

In the literature of probability theory, the characteristic function of a probability distribution is nothing more nor less than its Fourier transform. If, then, we write the probability distribution $P(X)$ and its characteristic function $\bar{P}(\omega)$, the relationship connecting the two is just

$$\bar{P}(\omega) = \int_{-\infty}^{\infty} P(X) e^{-i\omega X} \, dX \tag{2.70}$$

with the inversion relationship applying just as is written in Eq. (2.11).

There is a close relationship between the characteristic function and the positive moments of its distribution function. This relationship can be demonstrated by expanding the exponential in Eq. (2.70) in a Taylor series:

$$\bar{P}(\omega) = \int_{-\infty}^{\infty} P(X) \left[1 - i\omega X + \frac{1}{2!} (i\omega X)^2 - \cdots \right] dX \tag{2.71}$$

which when integrated gives

$$\bar{P}(\omega) = 1 - \frac{1}{2!} \langle X^2 \rangle \omega^2 + \frac{1}{4!} \langle X^4 \rangle \omega^4 - \frac{1}{6!} \langle X^6 \rangle \omega^6 + \cdots$$
$$- i \left[\langle X \rangle \omega - \frac{1}{3!} \langle X^3 \rangle \omega^3 + \frac{1}{5!} \langle X^5 \rangle \omega^5 - \cdots \right] \tag{2.72}$$

In principle, if all of the moments of $P(X)$ are known and finite, $\bar{P}(\omega)$ may be written as a Taylor series in ω, and $P(X)$ can be obtained from Eq. (2.11). Or, one could say that specifying all of the moments of a distribution function totally determines the form of $P(X)$. In practice, of course, all of the moments are never known, and even if they were, the series form of $P(X)$ obtained by integration of Eq. (2.72) would usually be too slowly convergent to be of any value. It can be shown, however, that many of the most important features of distribution functions can be determined from their first few moments.

Before we leave the treatment of moment expansions, we will take a look at them from a somewhat different perspective—one which will prove to be useful in the next section. We will begin by writing Eq. (2.10) as

$$\bar{f}(\omega) = \int_{-\infty}^{\infty} f(x)[\cos(\omega x) - i \sin(\omega x)] \, dx \qquad (2.73)$$

Since $f(x)$ is a probability distribution, it is a real function. This, and the fact that the cosine and sine are even and odd functions, respectively, make it clear that the only effect of changing the sign of ω will be a change in the sign of the imaginary part of $\bar{f}(\omega)$. This piece of information is most useful if we cast the characteristic function in the following, somewhat different, form:

$$\bar{f}(\omega) = A(\omega)e^{i\Phi(\omega)}$$
$$= A(\omega)\{\cos[\Phi(\omega)] + i \sin[\Phi(\omega)]\} \qquad (2.74)$$

where $A(\omega)$ is the absolute value of $\bar{f}(\omega)$ and $\Phi(\omega)$ is the phase angle. Examining Eq. (2.74) shows immediately that the only way that changing the sign of ω can result in changing the sign of the imaginary part of $\bar{f}(\omega)$ is if $A(\omega)$ is an even function and $\Phi(\omega)$ is an odd function. It is thus possible to write the two functions in Taylor series form as

$$A(\omega) = 1 + \frac{a_2}{2!} \omega^2 + \cdots \qquad (2.75)$$

and

$$\Phi(\omega) = b_1\omega + \frac{b_3}{3!} \omega^3 + \cdots \qquad (2.76)$$

Inserting the above two series into Eq. (2.74), we find

$$\bar{f}(\omega) = \left[1 + \frac{a_2}{2!} \omega^2 + \cdots\right]e^{i[b_1\omega + \cdots]}$$
$$= \left[1 + \frac{a_2}{2!} \omega^2 + \cdots\right]\left[1 + ib_1\omega + \cdots - \frac{b_1^2}{2!} \omega^2 + \cdots\right]$$
$$= 1 + ib_1\omega - \frac{1}{2!} (b_1^2 - a_2)\omega^2 + \cdots \qquad (2.77)$$

Comparing Eq. (2.77) with Eq. (2.72) shows that

$$b_1 = -\langle X \rangle \qquad \text{and} \qquad a_2 = \langle X \rangle^2 - \langle X^2 \rangle$$

$$= -\sigma^2 \qquad (2.78)$$

where σ^2 is called the variance of $f(X)$. Since it is given by the equation

$$\sigma^2 = \int_{-\infty}^{\infty} (\langle X^2 \rangle - X^2) f(X)\, dX \qquad (2.79)$$

it is a measure of the dispersion of $f(X)$.

2.3.5. Linear Combinations of Random Variables

There are many instances in science and engineering where a random variable of interest is the linear combination of independent random variables. For example we could write

$$X = x_1 + x_2 + x_3 + \cdots \qquad (2.80)$$

where X is the dependent and x_i the independent random variables. It is often the case that the distribution functions for all of the x_i are known, but it is the distribution function of X that is of interest. An example of this type of problem is encountered in measurement theory—the dependent random variable is the total error in a measurement, and the x_i are contributions from various sources. One is interested in the distribution of experimental errors given the distribution functions associated with the various contributing sources.

We will first consider a dependent random variable which is the sum of only two independent random variables:

$$X = x_1 + x_2 \qquad (2.81)$$

The probability of observing x_1 and x_2 in an infinitesimal element of area $dA = dx_1\, dx_2$ is given by

$$P_1(x_1)P_2(x_2)\, dA \qquad (2.82)$$

It follows, therefore, that the probability of observing x_1 and x_2 in any area A in the $x_1 - x_2$ plane is given by the integral

$$\int_A P_1(x_1)P_2(x_2)\, dA \qquad (2.83)$$

We will now consider that particular area which lies between the two straight lines,

$$x_1 + x_2 = X \quad \text{and} \quad x_1 + x_2 = X + \delta X \qquad (2.84)$$

For the moment, δX will be considered to be small but not infinitesimal. The probability of observing x_1 and x_2 simultaneously in this area is also the probability of finding the dependent random variable in the interval between X and $X + \delta X$, which is approximated by $P(X)\delta X$. The approximation becomes better as δX becomes smaller. For this example, the integral becomes

$$\int_{x_2=-\infty}^{\infty} \int_{x_1=X-x_2}^{X+\delta X-x_2} P_1(x_1)P_2(x_2)\,dx_1\,dx_2 \qquad (2.85)$$

If δX is sufficiently small, the above integral is approximated by

$$\int_{-\infty}^{\infty} P_1(X - x_2)\delta X P_2(x_2)\,dx_2 \qquad (2.86)$$

where the approximation becomes better as δX becomes smaller. Since the above integral approximates the probability of finding the dependent random variable in an interval between X and $X + \delta X$, the two expressions can be set equal to each other, giving

$$P(X)\delta X = \int_{-\infty}^{\infty} P_1(X - x_2)\delta X P_2(x_2)\,dx_2 \qquad (2.87)$$

We can imagine that we first cancel δX from both sides of Eq. (2.87) and then let δX approach zero. Since the above relationship becomes exact in the limit as $\delta X \to 0$, this process demonstrates that the probability distribution for X is just the convolution of the probability distributions for x_1 and x_2, i.e.,

$$P(X) = P_1(x_1) * P_2(x_2) \qquad (2.88)$$

This relationship can easily be extended to dependent random variables that are the sum of n independent random variables to show that

$$P(X) = P_1(x_1) * P_2(x_2) * \cdots * P_n(x_n) \qquad (2.89)$$

and it follows directly from Eq. (2.89) that the characteristic function of $P(X)$ is just the product of the characteristic functions of the $P_i(x_i)$.

Useful information about the first two moments of a dependent random variable can be obtained by combining the moment and convolution relationships. We first combine the results from Eqs. (2.89) and (2.72):

$$\bar{P}(\omega) = \prod \left[1 - i\langle x_i \rangle \omega - \frac{1}{2!} \langle x_i^2 \rangle \omega^2 + i\frac{1}{3!} \langle x_i^3 \rangle \omega^3 \right.$$

$$\left. + \frac{1}{4!} \langle x_i^4 \rangle \omega^4 - \cdots \right] \tag{2.90}$$

where the product is taken over the n independent random variables. Expanding the product gives

$$\bar{P}(\omega) = 1 - i\sum_{i=1}^{n} \langle x_i \rangle \omega - \left[\frac{1}{2!} \sum_{i=1}^{n} \langle x_i^2 \rangle + \sum_{i=1}^{n-1} \sum_{j=i+1}^{n} \langle x_i \rangle \langle x_j \rangle \right] \omega^2 \tag{2.91}$$

Comparing Eqs. (2.91) and (2.72) shows that

$$\langle X \rangle = \sum_{i=1}^{n} \langle x_i \rangle \tag{2.92}$$

and

$$\langle X^2 \rangle = \sum_{i=1}^{n} \langle x_i^2 \rangle - \sum_{i=1}^{n-1} \sum_{j=i+1}^{n} \langle x_i \rangle \langle x_j \rangle \tag{2.93}$$

Equations (2.92) and (2.93) are useful in determining the first and second moments of $f(X)$ when these two moments are known for all of the $f_i(x_i)$. What is more important for our purposes, however, is the application of these two equations in our discussion of the central limit theorem.

Before leaving the subject of the relationship between the moments of independent and dependent random variables, there is one more important relationship which should be considered—the relationship between means and variances. We will start by using the series form for characteristic functions as in Eq. (2.77). Using the representation gives

$$f(\omega) = \prod_{i=1}^{n} \left[1 - \frac{\sigma_i^2}{2!} \omega^2 + \cdots \right] e^{i[\langle x_i \rangle + \cdots]} \tag{2.94}$$

$$= \left[1 - \frac{\sum \sigma_i^2}{2!} \omega^2 + \cdots \right] e^{i[\sum \langle x_i \rangle \omega + \cdots]} \tag{2.95}$$

Comparing Eqs. (2.94) and (2.77) shows that

$$\sigma^2 = \sum \sigma_i^2 \quad \text{and} \quad \langle X \rangle = \sum \langle x_i \rangle \tag{2.96}$$

where the summations are taken over all of the independent random variables.

2.3.6. The Central Limit Theorem

Much more remains to be said about the behavior of the distribution functions of random variables that are the linear combinations of independent random variables. In particular, it can be shown that provided the distribution functions of the independent random variables satisfy some rather general criteria, the distribution function of the dependent random variable approaches a Gaussian function as the number of independent random variables approaches infinity. That is,

$$f(X) \rightarrow \frac{1}{\sigma(2\pi)^{1/2}} e^{-(X-\langle X \rangle)^2/2\sigma^2} \tag{2.97}$$

and

$$\bar{f}(\omega) \rightarrow e^{[(-\sigma^2\omega^2/2)-i\langle X \rangle \omega]} \qquad \text{as } n \rightarrow \infty \tag{2.98}$$

The proof of the central limit theorem is straightforward if we make two assumptions:

(1) All of the positive moments of the independent random variables are finite.

(2) $\sigma^2 = \sum \sigma_i^2 \rightarrow \infty$ as $n \rightarrow \infty$.

The first assumption is sufficient but not necessary. It should, however, pose no difficulty to the physical scientist or engineer. The *true* distribution functions associated with nearly all measurable (but random variable) quantities must be equal to zero outside of the range of possible measurable values. The *true* distribution is thus to be distinguished from mathematical approximations used to describe it over convenient ranges of values. If all of the positive moments are finite, the characteristic function of a distribution can always be expanded as a Taylor series as in Eq. (2.72).

The second assumption is, however, necessary, and we will see later in this chapter that this fact has implications with respect to the statistics of chainlike molecules.

We begin the analysis by taking the natural logarithm of $\bar{f}(\omega)$ as it is written in Eq. (2.61):

$$\ln[\bar{f}(\omega)] = \ln[\prod \bar{f}_i(\omega)] \tag{2.99}$$

which gives

$$\ln[\bar{f}(\omega)] = \sum \ln\left[1 - i\langle x_i \rangle \omega - \frac{\langle x_i^2 \rangle}{2!}\omega^2 + i\frac{\langle x_i^3 \rangle}{3!}\omega^3 + \frac{\langle x_i^4 \rangle}{4!}\omega^4 - \cdots\right] \tag{2.100}$$

We can consider the log terms to be of the form $\ln[1 + (\cdots)]$. If we then expand them in Taylor series in the functions $(\cdots)$ and regroup the terms in powers of ω, we get

$$\ln[\bar{f}(\omega)] = \sum \left\{ -i\langle x_i \rangle \omega - \left[\frac{\langle x_i \rangle^2}{2} - \frac{\langle x_i \rangle^2}{2} \right] \omega^2 \right.$$

$$+ i\left[\frac{\langle x_i^3 \rangle}{6} - \frac{\langle x_i \rangle \langle x_i^2 \rangle}{2} + \frac{\langle x_i \rangle^3}{3} \right] \omega^3$$

$$\left. + \left[\frac{\langle x_i \rangle^4}{24} - \frac{\langle x_i \rangle \langle x_i^3 \rangle}{6} + \frac{\langle x_i \rangle^2 \langle x_i^2 \rangle}{2} + \frac{\langle x_i^2 \rangle^2}{8} - \frac{\langle x_i \rangle^4}{4} \right] \omega^4 \cdots \right\}$$

$$\text{(2.101)}$$

From Eq. (2.92) we identify $\sum \langle x_i \rangle$ as the mean of $f(X)$, $\langle X \rangle$. Although it is not so obvious, Eq. (2.78) and a little algebraic rearranging of terms show that

$$\sum \langle x_i^2 \rangle - \sum \langle x_i \rangle^2 = \langle X^2 \rangle - \langle X \rangle^2$$

$$= \sigma^2 \qquad \text{(2.102)}$$

where σ^2 is the variance of the distribution $f(X)$. Equation (2.101) can thus be rewritten

$$\ln[\bar{f}(\omega)] =$$

$$-i\langle X \rangle \omega \left\{ 1 - \frac{\sum [\langle x_i^3 \rangle - 3\langle x_i \rangle \langle x_i^2 \rangle + 2\langle x_i \rangle^3]}{6 \sum \langle x_i \rangle} \omega^2 + \cdots \right\} - \frac{\sigma^2 \omega^2}{2}$$

$$\times \left\{ 1 - \frac{\sum [\langle x_i^4 \rangle - 4\langle x_i \rangle \langle x_i^3 \rangle + 12\langle x_i \rangle^2 \langle x_i^2 \rangle + 3\langle x_i^2 \rangle^2 - 6\langle x_i \rangle^4]}{12 \sum [\langle x_i^2 \rangle - \langle x_i \rangle^2]} \omega^2 + \cdots \right\}$$

$$\text{(2.103)}$$

One of the assumptions introduced in the development was that all of the moments remain bounded. It follows, therefore, that all of the large fractions in Eq. (2.103) remain bounded as $n \to \infty$.

We can thus write $\bar{f}(\omega)$ in the following form:

$$\bar{f}(\omega) = \exp[-(\sigma \omega^2/2)(1 + A_2 \omega^2 + \cdots) - i\langle X \rangle \omega(1 + B_2 \omega^2 + \cdots)] \quad \text{(2.104)}$$

where the coefficients A_i and B_i are finite no matter how large n may be. We note, however, that σ can be made as large as desired simply by taking n to be sufficiently large. It is thus possible to take n sufficiently large that the exponential in Eq. (2.104) can be made as near zero as we desire before the two power series in parentheses in Eq. (2.104) differ from unity. In other words, the right-hand side of Eq. (2.104) can always be made as close to the right-hand side of Eq. (2.98) as desired by taking n sufficiently large. The central limit theorem has thus been proven.

3. THE CLASSICAL RANDOM-FLIGHT MODEL

Once it was recognized that the structural feature that distinguishes polymers from their small-molecule counterparts is their long chainlike structure, an effort was mounted to explain their unique physical–chemical properties in terms of that structure.[12] Rotation about the backbone bonds of flexible polymer molecules gives them the ability to take on an almost unlimited variety of shapes, or conformations. It was therefore recognized that such theories would be statistical in nature. What was thus required was a simple model that would demonstrate the statistical nature of an assembly of flexible, linear polymer molecules, as well as contain information about their fully extended length (and by implication, the molecular weight of the molecules they are intended to represent).

The simplest model that offers these features is a chain of backbone bonds connected by completely flexible joints. This is indeed the first model that was used in the statistical mechanics of polymer molecules, and in spite of its simplicity, it went a long way toward explaining a variety of polymer properties. It is this model that we will examine in detail in this section. We will consider the statistical features of this model from the point of view of probability theory, with special emphasis on the application of characteristic functions. Discussions of the original work, which used a variety of mathematical approaches, are given in a number of treatises.[1,3,13]

Since this model is only intended to give a first approximation to the statistics of a collection of polymer molecules, the backbone bonds of the contiguous set are considered to be simple directed line segments, or vectors, with no associated volume or mass. It is important to keep in mind that these "backbone bonds" are not to be considered the chemical bonds that make up the backbone of a real polymer molecule, but only as a mathematical construct that is introduced in the hope that it will demonstrate conformational statistics similar to those of a collection of polymer molecules.

The vectors spanning the end-to-end distances of members of an assembly of such completely flexible chains and their orthogonal components were the first measures of the spatial dimensions of chainlike molecules to be examined in great detail. For mathematical simplicity, we will consider the backbone bonds as a set of n vectors, with l_i as the ith member of a set. Every vector of the set is of length l. The end-to-end distance vector $\mathbf{L}$ of any member of an assembly of such chains is then given by

$$\mathbf{L} = \sum l_i \qquad (i = 1 \text{ to } n) \tag{3.1}$$

and the x, y, and z components of $\mathbf{L}$ are given by

$$X = \sum x_i, \qquad Y = \sum y_i, \qquad Z = \sum z_i, \qquad (i = 1 \text{ to } n) \qquad (3.2)$$

where x_i, y_i, and z_i are the orthogonal components of the corresponding backbone bond vectors.

3.1. Random-Flight Chains: The Discrete Model

The conformational statistics of an assembly of completely flexible chains can be viewed in terms of the random flight problem. This classical problem in physics can, however, be considered from both the discrete and continuous points of view. Although the continuous picture is more in line with reality, the discrete picture is often drawn upon to explain the statistical behavior of polymer chains. We will thus start from the discrete model of the random flight and then move into the continuous picture.

In the discrete random flight problem, we will imagine that steps are taken from point to nearest-neighbor point on a lattice. For our purposes, we will consider only square lattices. Each step will thus be to one of the six nearest-neighbor points, and the probability of moving in any one of the six directions is equal. It is also assumed that there is no restriction on the number of times any lattice point can be visited. One aspect of the random-flight problem is determining the distribution of distances between the starting and finishing points for a large number of random walks of n steps.

This is a three-dimensional problem, but since the steps are taken at random, it can be considered as the superposition of three one-dimensional problems. We will thus start by considering the processes of stepping back and forth at random along a line with a step length of l. In this one-dimensional *random walk*, it is equally probable that any step will be in the positive or negative direction. If x_i is the displacement of the ith step, the probability function for x_i, $P(x_i)$, is given by the expression

$$P(x_i) = \tfrac{1}{2}[\delta(x_i - l) + \delta(x_i + l)] \qquad (3.3)$$

and it follows that the characteristic function of $P(x_i)$ is written

$$\bar{P}(\omega) = \tfrac{1}{2}[e^{-i\omega l} + e^{+i\omega l}] \qquad (3.4)$$

The total displacement after n' steps, X, is thus given by

$$X = \sum x_i \qquad (3.5)$$

The development in Section 2.3.5 indicates that the characteristic function of the probability distribution for X, $P_{n'}(X)$, is the product of n' functions of the type given in Eq. (3.4). The result is

$$\bar{P}_{n'}(\omega) = \frac{1}{2^{n'}} [e^{-i\omega l} + e^{+i\omega l}]^{n'} \tag{3.6}$$

or

$$\bar{P}_{n'}(\omega) = \frac{e^{-in'l\omega}}{2^{n'}} [1 + e^{+2i\omega l}]^{n'} \tag{3.7}$$

Expanding the term in the square bracket gives

$$\bar{P}_{n'}(\omega) = \frac{e^{-in'l\omega}}{2^{n'}} \sum_{m=0}^{n'} \frac{n'!\, e^{2ilm\omega}}{(n'-m)!\,m!} \tag{3.8}$$

or

$$\bar{P}_{n'}(\omega) = \frac{1}{2^{n'}} \sum_{m=0}^{n'} \frac{n'!\, e^{il\omega(2m-n')}}{(n'-m)!\,m!} \tag{3.9}$$

Since the characteristic function of $P_{n'}(X)$ is given by Eq. (3.9), the distribution function itself is given by

$$P_{n'}(X) = \frac{1}{2^{n'}} \sum_{m=0}^{n'} \frac{n'!}{(n'-m)!\,m!}\, \delta[X - l(2m - n')] \tag{3.10}$$

The distribution function in Eq. (3.10) is a set of delta functions symmetrically distributed about $X = 0$. Aside from the factor $(1/2^{n'})$, the coefficients of the delta functions are the binomial coefficients. For a walk of 11 steps they are equal to

$$1\text{-}11\text{-}55\text{-}165\text{-}330\text{-}462\text{-}462\text{-}330\text{-}165\text{-}55\text{-}11\text{-}1$$

The coefficients of the delta functions are plotted in Figure 1 along with a smooth curve which represents the envelope of the points. The envelope is a typical bell-shaped curve. If the number of steps increases without bound, the points on the x axis get closer together and the distribution function goes from a discrete distribution to a continuous one. If n' is sufficiently large, the distribution is well approximated by a continuous function. Indeed a simple closed analytical form can be derived for that distribution. We begin the derivation by noting the following one-to-one correspondence between the values of X and m:

$$m = \frac{X}{2l} + \frac{n'}{2} \tag{3.11}$$

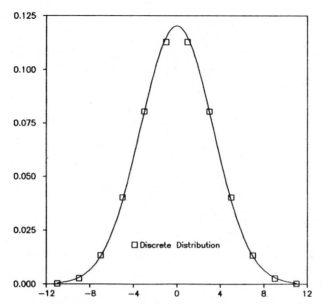

FIGURE 1. A comparison between the discrete distribution function and its Gaussian approximation for a one-dimensional random walk of 11 steps. The ordinate units for the Gaussian distribution are arbitrary.

The binomial coefficients can thus be written

$$\text{Coef.} = \frac{n'!}{\left(\dfrac{n'}{2} - \dfrac{X}{2l}\right)!\left(\dfrac{n'}{2} + \dfrac{X}{2l}\right)!} \tag{3.12}$$

Stirling's approximation, $\ln(N)! \approx N\ln(N) - N$ (where N is any large number), can be used to approximate the factorials in Eq. (3.12). The approximation becomes better with increasing n'. We note that the approximation fails as $X \to n'l$, the length of the fully extended chain. As long as $(X/n'l)$ is sufficiently small, however, the coefficients are adequately approximated by the expression

$$\ln(\text{Coef.}) = n'\ln(2) \frac{n'}{2}\left[\left(\frac{X}{n'l}\right)^2 + \frac{1}{6}\left(\frac{X}{n'l}\right)^4 + \cdots\right] \tag{3.13}$$

As long as we are concerned with sufficiently small values of $(X/n'l)$, terms beyond the second order in the square brackets are negligible, and the result is

$$\text{Coef.} = 2^{n'} e^{-X^2/2n'l^2} \tag{3.14}$$

Aside from the factor $(1/2^{n'})$, Eq. (3.14) gives the envelope of the coefficients of the delta functions in Eq. (3.10). For sufficiently large n', the discrete distribution can thus be adequately approximated by a continuous distribution which is proportional to the right-hand side of Eq. (3.13). We note, however, that the exponential part is the only part that is a function of X. It is thus possible to write down the normalized continuous distribution function which approximates the sum of delta functions given in Eq. (3.10). The result is

$$P_{n'}'(X) = [1/l(2n'\pi)^{1/2}]e^{-(1/2n'l^2)X^2} \tag{3.15}$$

We must recall, however, that n' is the number of steps taken along a directed line. This model thus applies to one dimension out of a three-dimensional random flight. To apply it to the three-dimensional problem, we must make one additional assumption. It will be assumed that if a total of n steps are taken, there will be $(n/3)$ steps taken in each of the three orthogonal directions. With that assumption we can write

$$n' = n/3 \tag{3.16}$$

and rewrite Eq. (3.15) as

$$P_n(X) = [\sqrt{3}/l(2n\pi)^{1/2}]e^{-(3/2nl^2)X^2} \tag{3.17}$$

The second moment of the above distribution is

$$\langle X^2 \rangle = \tfrac{1}{3}nl^2 \tag{3.18}$$

Since

$$L^2 = X^2 + Y^2 + Z^2$$
$$\langle L^2 \rangle = \langle X^2 \rangle + \langle Y^2 \rangle + \langle Z^2 \rangle \tag{3.19}$$

and since all three directions are equivalent, it follows that

$$\langle L^2 \rangle = nl^2 \tag{3.20}$$

3.2. Random-Flight Chains: The Continuous Model

Although we will see that the equations derived from the discrete random flight model are correct, the discrete nature of the model is unsatisfying. A model in which the backbone vectors can take on any orientation is intuitively more appealing, so we will turn our attention in that direction.

Before investigating the statistics of completely flexible chains in detail, it is useful to consider their mean-square end-to-end distance $\langle L^2 \rangle$. Since $L^2 = \mathbf{L} \cdot \mathbf{L}$ we can write

$$L^2 = [\textstyle\sum l_i] \cdot [\sum l_j]$$
$$= \textstyle\sum l_i^2 + \text{cross terms of the form } l_i \cdot l_j \text{ where } i \neq j \quad (3.21)$$

Since l_i^2 is the square of the length of the ith backbone bond, and since each bond is of the same length, l, we find that

$$L^2 = nl^2 + \text{cross terms} \quad (3.22)$$

Taking the mean value of both sides of Eq. (3.22) gives

$$\langle L^2 \rangle = nl^2 + \langle \text{cross terms} \rangle \quad (3.23)$$

Since each cross term $l_i \cdot l_j$ is equal to $l^2 \cos \theta$, where θ is the angle between vectors l_i and l_j, we can write

$$\langle l_i \cdot l_j \rangle = l^2 \int_0^{2\pi} W_\theta(\theta) \cos \theta \, d\theta \quad (3.24)$$

where $W_\theta(\theta)$ is the distribution function characterizing θ. Since the chains are completely flexible, it is equally likely to find θ within any interval $\delta\theta$ that lies within the range of $\theta = 0$ to 2π. It follows, therefore, that $W_\theta(\theta)$ must be constant over this interval, so the normalized distribution function is given by

$$W_\theta(\theta) = (1/2\pi) \quad \text{for } 0 \leq \theta \leq 2\pi$$
$$= 0 \quad \text{otherwise} \quad (3.25)$$

Substituting this expression for $W_\theta(\theta)$ into Eq. (3.24) and integrating demonstrates that every $\langle \text{cross term} \rangle$ is equal to zero.

As long as the backbone vectors are completely independent, it thus follows that

$$\langle L^2 \rangle = nl^2 \quad (3.26)$$

which is the same expression [Eq. (3.20)] derived using the discrete random flight model.

3.3. The Orthogonal Components of the End-to-End Distance Vector

The relationship between the mean-square end-to-end distance vector and the mean-square values of its orthogonal components can easily be determined by starting with the relationship

$$L^2 = X^2 + Y^2 + Z^2 \tag{3.27}$$

It follows from Eq. (3.27) that

$$\langle L^2 \rangle = \langle X^2 \rangle + \langle Y^2 \rangle + \langle Z^2 \rangle \tag{3.28}$$

Since the three orthogonal directions are equivalent, the mean square of the three components must be equal, with the result that

$$\langle X^2 \rangle = \langle Y^2 \rangle = \langle Z^2 \rangle = (1/3)nl^2 \tag{3.29}$$

It is easy to derive Eq. (3.29) from the distribution functions for the orthogonal components of the backbone vectors themselves. Consider, for example, the distribution of x_i, the x component of the ith backbone vector. Since l_i can be oriented at random to the x axis, the projection of l_i is also random between the limits of $-l$ to $+l$. The random variable x_i is thus characterized by a uniform distribution over the range of $-l$ to $+l$, i.e.,

$$W_x(x_i) = 1/2l \qquad \text{for } -l \le x_i \le +l$$
$$= 0 \qquad \text{otherwise} \tag{3.30}$$

The mean-square value of x_i can thus be written

$$\langle x_i^2 \rangle = \int_{-l}^{+l} \frac{x_i^2 \, dx_i}{2l} = \frac{1}{3} l^2 \tag{3.31}$$

From Eq. (3.2) we find

$$\langle X^2 \rangle = \sum \langle x_i^2 \rangle \tag{3.32}$$

Since there are n terms in the above summation, and since all of the $\langle x_i^2 \rangle$ are equal to $(1/3)l^2$, it follows that

$$\langle X^2 \rangle = (1/3)nl^2 \tag{3.33}$$

which, of course, is identical to Eq. (3.29).

The treatment that gives the mean-square values of the end-to-end distance vector and its orthogonal components of random flight chains is quite straightforward. The methods for determining distribution functions of L, X, Y, and Z are, however, a bit more involved. We will turn our attention first to the orthogonal components. Equations (3.2) show that X, Y, and Z are dependent random variables that are sums of the x_i, y_i, or z_i. Since each of the independent random variables is characterized by the same distribution function, they have the same mean (zero) and the same second moment ($l^2/3$). The conditions are thus obeyed for the central limit theorem to apply. We recall that the central limit theorem states that if the distribution functions of the independent random variables are suitably well behaved (and in this case they are), the distribution function of the dependent random variable approaches a Gaussian distribution with increasing number of independent random variables. For sufficiently large n, we can thus write

$$W_X(X) = (\beta/\sqrt{\pi})e^{-\beta^2 X^2} \tag{3.34}$$

where

$$\beta^2 = (1/2\langle X^2 \rangle) = (3/2nl^2) \tag{3.35}$$

with identical relationships for the y and z components.

An important question is, of course, how large n must be before Eq. (3.34) is obeyed well enough for most practical purposes. The answer to this question has been explored extensively, including analyses of deviations from Gaussian behavior at high extensions.[3,13] The approaches have been ingenious and varied, and represent some of the earliest studies of conformational statistics. In retrospect, however, we find that much of what was learned can be derived directly from the characteristic function of $W_X(X)$.

Equation (3.30) gives the probability distribution for the x component of a single step of a random flight of n steps. The characteristic function of this distribution is presented in Eq. (2.30). Since the probability distribution for X is the nth-order convolution for the distributions of the x_i [see Eqs. (2.61) and (2.89)], the characteristic function of $W_X(X)$ is given by

$$\bar{W}_X(\omega) = \left[\frac{\sin(\omega l)}{\omega l}\right]^n \tag{3.36}$$

and

$$W_X(X) = \frac{1}{2\pi} \int_{-\infty}^{\infty} \left[\frac{\sin(\omega l)}{\omega l}\right]^n e^{i\omega X}\, d\omega \tag{3.37}$$

Since

$$\sin(\omega l) = \frac{1}{2i}[e^{+i\omega l} - e^{-i\omega l}] \tag{3.38}$$

Eq. (3.37) can be recast in the form

$$W_x(X) = \frac{1}{2\pi(2i)^n} \sum_{p=0}^{n} \int_{-\infty}^{\infty} (-1)^n \frac{n!\, e^{i[(X/l)+n-2p]\omega l}}{p!(n-p)!(\omega l)^n}\, d\omega \qquad (3.39)$$

The integrals in Eq. (3.39) are of the form

$$\int_{-\infty}^{\infty} \frac{e^{ib\xi}}{\xi^n}\, d\xi \qquad (3.40a)$$

$$= \frac{2\pi(i)^n}{(n-1)!}\, b^{n-1} \qquad \text{if } b \leq 0$$

$$= 0 \qquad \text{if } b > 0 \qquad (3.40b)$$

The function $W_x(X)$ is thus given exactly by the expression

$$W_x(X) = \frac{1}{2^n(n-1)!} \sum_{p=0}^{n} (-1)^n \frac{n!}{lp!(n-p)!}\, F_n(X) \qquad (3.41)$$

where

$$F_n(X) = [(X/l) + n - 2p]^{n-1} \qquad \text{if } X \leq l(2p-n)$$

$$= 0 \qquad \text{if } X > l(2p-n) \qquad (3.42)$$

For the case where $n = 3$, $W_x(X)$ is written in the following simple form:

$$lW_x(X) = \tfrac{9}{16} - \tfrac{3}{8}(X/l) + \tfrac{1}{16}(X/l)^2 \qquad \text{if } 1 \leq (X/l) \leq 3$$

$$= \tfrac{3}{8} - \tfrac{1}{8}(X/l)^2 \qquad \text{if } -1 \leq (X/l) \leq 1$$

$$= \tfrac{9}{16} + \tfrac{3}{8}(X/l) + \tfrac{1}{16}(X/l)^2 \qquad \text{if } -1 \leq (X/l) \geq -1$$

$$= 0 \qquad \text{otherwise} \qquad (3.43)$$

Figures 2 through 4 compare the exact distributions $W_x(X)$ for $n = 3$, 10, and 25, respectively, with the Gaussian approximation given in Eq. (3.34) over the range of dimensionless x component of the end-to-end distance $(X/\langle X^2 \rangle^{1/2}) = (X\sqrt{3}/\sqrt{n}l)$ from -4 to $+4$. The Gaussian approximation is surprisingly good even for a chain of only three links, and, except for near the origin, the Gaussian approximation is indistinguishable (on the scale of Figure 4) from the exact distribution. It should be stressed, however, that the exact distribution is always equal to zero for all $|X| > nl$, whereas the Gaussian approximation has nonzero values for all values of X. This is clearly demonstrated in Figure 2 for the case of $n = 3$. In practice, however, the mathematical probability represented by the area

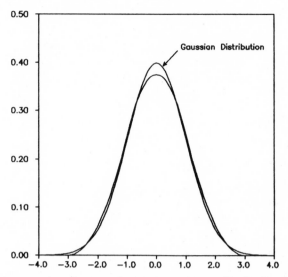

FIGURE 2. A comparison between the exact, continuous distribution function and its Gaussian approximation for a one-dimensional random walk of three steps.

under the nonphysical tails of the Gaussian distribution is insignificant, and errors are only significant for situations where one must consider chains at high extensions.

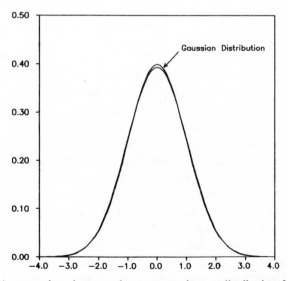

FIGURE 3. A comparison between the exact, continuous distribution function and its Gaussian approximation for a one-dimensional random walk of ten steps.

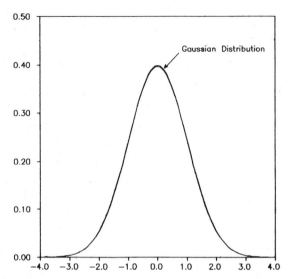

FIGURE 4. A comparison between the exact, continuous distribution function and its Gaussian approximation for a one-dimensional random walk of 25 steps. The difference is not detectable on the scale of this figure.

4. THE GENERALIZED RANDOM-FLIGHT MODEL: LINEAR AND BRANCHED CHAINS

The random flight model described in considerable detail in the previous section serves well as a first approximation to the conformational statistics of flexible polymer molecules. Nonetheless, its limitations are clear, and there has been continuing effort to develop more realistic models of chainlike macromolecules. The most obvious extension of the random flight model is the incorporation of methods for dealing with chemical structure at the repeat unit level. These effects are defined as *short range*.

The *rotational-isomeric* technique is a computational protocol for generating conformational statistics that deals explicitly with short-range effects. This method begins with the random-flight model. In this case, however, the bond vectors are, in fact, considered to be the chemical bonds making up the chain backbone, but they are not assumed to be completely flexible. Indeed, this treatment accounts for fixed bond angles and potential energies of rotation about backbone bonds in the absence of solvent molecules; considering the effects of solvent is yet to come. The rotational-isomeric treatment of short-range effects—at least in the absence of solvent—has become a well-developed art. That material is, however, well beyond the scope of this chapter, and will not be considered further here. Early work was reviewed by Volkenstein[13] and considerable more detail and newer material is covered in *Statistical Mechanics of Chain Molecules* by Flory.[14]

Secondly, even if short-range effects are adequately accounted for, one must still consider the effects on chain statistics of conformations for which the chain "bends back upon itself" at one or more places with concomitant interactions between repeat units. Since these repeat units are spaced far apart along the chain backbone, their interactions are defined as long-range effects (even though the effect accounts for contact between repeat units). As detailed as the rotational-isomeric technique is, it does not deal with these effects. Long-range effects will, however, be considered later in this chapter, where it will be pointed out that short-range and long-range effects are dealt with in an additive fashion.

In this section, we present an alternative to the model discussed in Section 3. Although it would appear that this second approach completely side-steps the issue of short-range effects, we will discuss how that notion enters in. More importantly, however, this alternative to the simpler random flight model is far more convenient in dealing with long-range effects and, as we will see in Sections 6 and 7, with chain dynamics.

The model we are going to adopt is one in which the detailed structure of the chainlike molecule is ignored. To this level of approximation, the molecules are replaced by an assembly of chains, each of which can be thought of as a string of contiguous point masses. The masses are often pictorially represented as beads. The model is further defined by stipulating that the relative positions of adjacent point masses are distributed according to the Gaussian probability density, i.e.,

$$W(x_i) = (\sqrt{n}\,\beta/\sqrt{\pi}\,)\exp(-n\beta^2 x_i^2) \qquad (4.1)$$

with analogous expressions for the distributions of y_i and z_i. Comment on why the Gaussian parameter is often written as $\sqrt{n}\beta$ will be given shortly.

Polymer–polymer and polymer–solvent interactions (i.e., long-range effects) are then incorporated into the model by defining an effective pair-wise-additive potential of interaction between beads.

This model can be justified only on the basis of how well it describes a wide range of polymer behavior, and it has indeed served well in that capacity. It is useful, however, to examine in detail the question of how this highly idealized model is related to the molecular structure of real polymer molecules, and to explore the circumstances under which the model

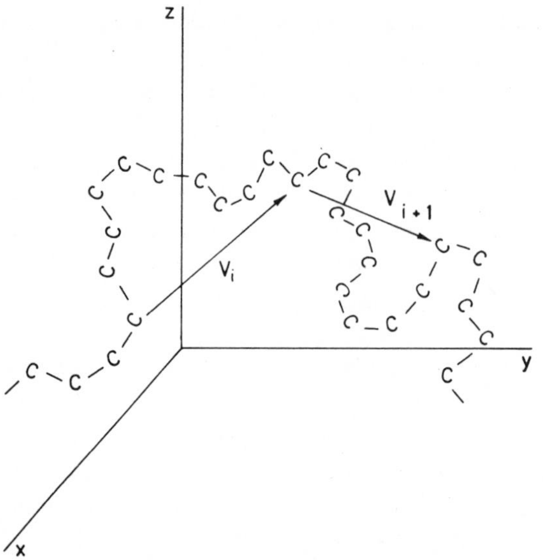

FIGURE 5. Statistical segments i and $i + 1$ for a polymer molecule with a carbon backbone. Only the backbone carbon atoms are shown.

can be expected to fail. Consider a molecule of a vinyl polymer placed in a Cartesian coordinate system as shown in Figure 5. For simplicity, only the carbon atoms are shown. Imagine that we define a set of vectors connecting every ten carbon atoms (one for every five repeat units). Let the ith vector $\mathbf{V}_i$ be fixed. Vector $\mathbf{V}_{i+1}$ can then be varied over a range of lengths and directions by rotating carbon atoms about the carbon–carbon bonds bridged by $\mathbf{V}_i$ and $\mathbf{V}_{i+1}$, consistent with the fixed length and direction of $\mathbf{V}_i$. Given the potential functions for rotation about all the carbon–carbon bonds bridged by $\mathbf{V}_i$ and $\mathbf{V}_{i+1}$, it is, in principle, possible to calculate the distribution functions for the three orthogonal components of $\mathbf{V}_{i+1}$. If these distribution functions were to be independent of the length and orientation of $\mathbf{V}_i$, $\mathbf{V}_{i+1}$ would be said to be stochastically independent of $\mathbf{V}_i$. In this example, the vectors bridge only ten carbon–carbon bonds, so $\mathbf{V}_{i+1}$ would not be expected to be quite stochastically independent of $\mathbf{V}_i$. The practical issue is, of course, whether or not $\mathbf{V}_{i+1}$ would be *sufficiently* stochastically independent of $\mathbf{V}_i$ to make the assumption of independence a reasonable one when formulating a model for flexible polymer molecules.

It is clear that if the vectors spanned more than ten backbone bonds they would be more nearly stochastically independent of one another. It is true, however, that rigorous stochastic independence is approached only as the number of backbone bonds spanned by the vectors approaches infinity. Nonetheless, it is assumed that there is some reasonable number of backbone bonds that can be spanned by vectors such that they are sufficiently stochastically independent for most practical purposes. After it has been extablished that a set of vectors is sufficiently stochastically independent (according to some criterion that will not be elaborated upon here), it is, at least in principle, possible to examine the distribution functions characterizing their orthogonal components. If the distribution functions are Gaussian—or at least sufficiently so for practical purposes—the set of vectors will represent the statistical segments that are used to model the conformational features of flexible polymer molecules.

If, however, the set of stochastically independent vectors do not have components with distribution functions that are "sufficiently Gaussian," one can define a new set of vectors, each consisting of the sum of two or three or more contiguous members of the original set. Probability theory then guarantees that the components of the new vectors will be distributed according to functions that are better approximated by Gaussian distributions than the components of the original set. Indeed the central limit theorem guarantees that Gaussian behavior is approached rapidly with increasing numbers of original vectors taken for each vector in a new set.

The above arguments imply that we can always define a set of vectors, each bridging the same number of backbone bonds, that, for practical purposes, are stochastically independent and with orthogonal components that are distributed (except in the case of high extension) according to a Gaussian distribution. Admittedly, the above arguments are qualitative; only parts have ever been tested numerically. Nonetheless, they offer some insight into why the concept of the statistical segment has proven to be so useful in describing the conformational statistics of flexible, chainlike molecules.

The second idealization introduced in defining this so-called "spring-bead" model is the following: It is assumed that each molecular chain behaves as if its mass were localized in $n + 1$ equal units at both ends and at the head–tail junctions of the statistical segments, and that all polymer–polymer and polymer–solvent interactions can be described by a pairwise additive potential acting between the point masses (more will be said about this later in this chapter). In experiments designed to measure molecular phenomena in which the length scale is much greater than, say, the mean-square end-to-end distance of the statistical segments, assuming that the mass is distributed in this manner is clearly adequate; we can cite as an example the success of the classical light scattering method for determining the mean-square radius of gyration of polymer chains. The model requires

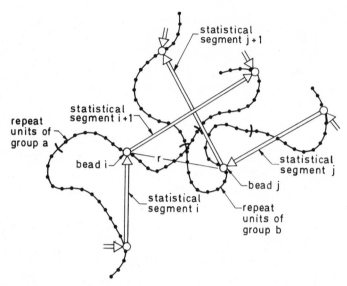

FIGURE 6. Groups of repeat units associated with beads *i* and *j* in the random–flight model.

careful inspection, however, when we consider the interpretation of the pairwise additive potential between the "beads."

Consider two beads *i* and *j* a distance *r* apart. The mean energy of interaction represents an average of the sum of interactions between all the nonbonded atoms associated with the two interacting beads. For example, the potential of interaction between beads *i* and *j* in Figure 6 is the sum of interaction energies of the chain atoms of group *a* with those of group *b*, averaged over all allowed positions of all the atoms included in statistical segments *i*, *i* + 1, *j*, and *j* + 1 consistent with fixed *r*. It is important to remember that the average potential is defined in this way whether or not beads *i* and *j* are part of the same molecule or different molecules.

For a given polymer, the bead–bead potential function is determined by the solvent and temperature. To the level of approximation of polymer behavior that this model offers, however, theory predicts that this potential must be a constant (independent of *r*) for bulk polymer. This is clearly the case since varying *r* has no effect on this average potential; none of the repeat units change their local environments as statistical segments *i*, *i* + 1, *j*, and *j* + 1 take on all values consistent with fixed *r*. The same can be said for polymer in sufficiently concentrated solution. More will be said about this important feature of the model later in this chapter.

4.1. Random-Flight Chains and Matrices: Determination of Mean Values

Figure 7 shows a model of a linear flexible polymer molecule where the vectors making up the backbone represent the statistical segments discussed in the last section. In this initial presentation we will consider only linear molecules; it will be shown later in this chapter how the analysis of conformational statistics can be expanded to include branched molecules. For the time being, we will assume that there is no energy of interaction between the beads. Although this model would seem to be unrealistic, we will find that it has relevance in real polymer–solvent systems.

4.1.1. The End-to-End Distance Vector

The first measures of the spatial dimensions of chainlike molecules that were examined in great detail were the end-to-end distance vector and its orthogonal components. If we define l_i as the ith member of the n statistical segment vectors, the end-to-end distance vector $\mathbf{L}$ and its x, y, and z components are given by

$$\mathbf{L} = \sum l_i \qquad (i = 1 \text{ to } n)$$
$$X = \sum x_i, \qquad Y = \sum y_i, \qquad Z = \sum z_i \qquad (i = 1 \text{ to } n) \tag{4.2}$$

The distributions of the x_i, y_i, and z_i are all Gaussian with the same distribution parameter $\sqrt{n}\beta$. By applying Eqs. (4.2), (2.89), and (2.62), it can be shown that the distributions functions for X, Y, and Z are also Gaussian with a parameter β, i.e.,

$$W_X(X) = (\beta/\sqrt{\pi}) \exp(-\beta^2 X^2) \tag{4.3}$$

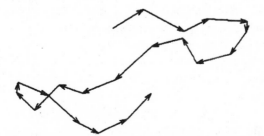

FIGURE 7. A chain of vectors representing the statistical segments in the random-flight model.

with analogous expressions for the distributions of Y and Z. Since the number of statistical segments is, within limits, arbitrary, and since it is the assembly of molecules that is the object of experimentation, it is reasonable to associate a fundamental parameter β with the molecular distributions, as was suggested by Flory,[1] and not with the distributions associated with the statistical segments.

4.1.2. Radius of Gyration

Although the statistics of the end-to-end distance vector and its components was worked out in meticulous mathematical detail,[3,13] this was more because the mathematics were manageable than because the end-to-end distance vector was of any intrinsic interest. It is, in fact, the radius of gyration which is the most important parameter in relating the conformational features of an assembly of polymer molecules to the physical properties of the material or its solutions. It is nonetheless true, however, that, for sufficiently long, unperturbed macromolecular chains, $\langle L^2 \rangle_0 = 6\langle S^2 \rangle_0$. This relationship does, then, give more than a little impetus for the study of the statistics of $\mathbf{L}$ and its components.

Before examining the statistics of the radius of gyration, we must remember an important feature of the model as it was defined in the previous section. It is assumed that there are few enough repeat units per statistical segment that each molecule behaves as if its mass were evenly divided among $n + 1$ masses located at the ends of the statistical segment vectors as shown in Figure 7. The reason for this assumption has to do with applications of the model to rheological problems and problems involving the interaction of the polymer molecules with radiation. In the case of rheological problems, we imagine the chainlike molecules to be moving in a continuum and demonstrating a viscous drag. The problem becomes manageable only if we can assume that the drag is applied at equally spaced centers along the chain backbone, i.e., that the drag acts at the center of the beads. In the case of macromolecules interacting with radiation, we assume that the scattering centers are the beads. It is clear that both of these assumptions would suffer greatly if too much of the mass was distributed along the distances connecting the heads and tails of the statistical segment vectors.

In summary, the model can be successful only to the extent that we can take enough repeat units per statistical segment to assure Gaussian behavior, but not too many to violate the notion that the mass can be assumed to be localized at the ends of the statistical segments.

Since we have assumed that the mass of the macromolecule is divided evenly between the $n + 1$ beads, the following expression can be written for the radius of gyration, S:

$$S^2 = (n + 1)^{-1} \sum \hat{r}_i^2 \tag{4.4}$$

where $\hat{r}_i$ is the distance of the ith mass from the center of mass. If we take $\hat{x}_i$, $\hat{y}_i$, and $\hat{z}_i$ as the three orthogonal components of $\hat{r}_i$, we can then define three corresponding orthogonal components of S by the equations

$$S_x^2 = (n + 1)^{-1} \sum \hat{x}_i^2$$
$$S_y^2 = (n + 1)^{-1} \sum \hat{y}_i^2 \tag{4.5}$$
$$S_z^2 = (n + 1)^{-1} \sum \hat{z}_i^2 \qquad (i = 1 \text{ to } n + 1)$$

In the rest of the development we must keep in mind that the origin of the coordinate system is the center of mass, so that

$$\sum \hat{x}_i = 0, \qquad \sum \hat{y}_i = 0, \qquad \text{and} \qquad \sum \hat{z}_i = 0 \tag{4.6}$$

In order to develop the relationship between S and the l_i, we begin by considering the vectors shown in Figure 7. We write

$$\mathbf{L}_i = \sum_{j=1}^{i-1} l_{ij} \tag{4.7}$$

and

$$\hat{\mathbf{r}}_i = \hat{\mathbf{r}}_1 + \mathbf{L}_i \tag{4.8}$$

There is no loss in generality in considering only one component of the above vector equations since behavior in all three orthogonal directions is independent. The x component of Eq. (4.8) is

$$\hat{x}_i = \hat{x}_1 + \sum x_j \qquad (j = 1 \text{ to } i - 1) \tag{4.9}$$

or, in more detail,

$$\hat{x}_2 = \hat{x}_1 + x_1$$
$$\hat{x}_3 = \hat{x}_1 + x_1 + x_2$$
$$\hat{x}_4 = \hat{x}_1 + x_1 + x_2 + x_3 \tag{4.10}$$
$$\vdots$$
$$\hat{x}_n = \hat{x}_1 + x_1 + x_2 + \cdots + x_{n-1}$$

Equations (4.6) and (4.9) can be combined to give

$$2\hat{x}_2 + \hat{x}_3 + \hat{x}_4 + \ldots = x_1$$
$$\hat{x}_2 + 2\hat{x}_3 + \hat{x}_4 + \ldots = x_1 + x_2 \tag{4.11}$$
$$\hat{x}_2 + \hat{x}_3 + 2\hat{x}_4 + \ldots = x_1 + x_2 + x_3$$

which, in matrix form, is written

$$\begin{bmatrix} 2 & 1 & 1 & \ldots & 1 \\ 1 & 2 & 1 & \ldots & 1 \\ 1 & 1 & 2 & \ldots & 1 \\ \cdot & \cdot & \cdot & \ldots & \cdot \\ 1 & 1 & 1 & \ldots & 2 \end{bmatrix} \begin{bmatrix} \hat{x}_2 \\ \hat{x}_3 \\ \hat{x}_4 \\ \cdot \\ \hat{x}_{n+1} \end{bmatrix} = \begin{bmatrix} 1 & 0 & 0 & \ldots & 0 \\ 1 & 1 & 0 & \ldots & 0 \\ 1 & 1 & 1 & \ldots & 0 \\ \cdot & \cdot & \cdot & \ldots & \cdot \\ 1 & 1 & 1 & \ldots & 1 \end{bmatrix} \begin{bmatrix} x_1 \\ x_2 \\ x_3 \\ \cdot \\ x_n \end{bmatrix} \tag{4.12}$$

The reciprocal of the first matrix in Eq. (4.12) can be shown to be equal to

$$(n+1)^{-1} \begin{bmatrix} n & -1 & -1 & \ldots & -1 \\ -1 & n & -1 & \ldots & -1 \\ -1 & -1 & n & \ldots & -1 \\ \cdot & \cdot & \cdot & \ldots & \cdot \\ -1 & -1 & -1 & \ldots & n \end{bmatrix} \tag{4.13}$$

It is easy to show that if both sides of Eq. (4.12) are multiplied from the left by the above matrix, the result is

$$\begin{bmatrix} \hat{x}_2 \\ \hat{x}_3 \\ \hat{x}_4 \\ \cdot \\ \cdot \\ \hat{x}_{n+1} \end{bmatrix} = (n+1)^{-1} \begin{bmatrix} 1 & -(n-1) & -(n-2) & -(n-3) & \ldots & -1 \\ 1 & 2 & -(n-2) & -(n-3) & \ldots & -1 \\ 1 & 2 & 3 & -(n-3) & \ldots & -1 \\ \cdot & \cdot & \cdot & \cdot & \ldots & \cdot \\ 1 & 2 & 3 & 4 & \ldots & n \end{bmatrix} \begin{bmatrix} x_1 \\ x_2 \\ x_3 \\ \cdot \\ x_n \end{bmatrix} \tag{4.14}$$

     There is an important aspect of the dimensionality of this problem which should be stressed. Within the framework of the model we are working with, the conformation, or shape, of a macromolecule is completely defined by fixing the lengths and orientations of the statistical segments —or in the one-dimensional case being considered, the values of all of the x_i. There are n statistical segments, but $n+1$ position coordinates for the masses (i.e., beads). It follows, therefore, that in the formulation of any problem in conformational statistics (which, of course, deals with shape), we must eliminate one of the position coordinates through Eq. (4.6) before the problem can be formulated in terms of the $\hat{x}_i$ coordinates. Indeed, that is why the matrix equation given above contains only n of the $n + 1$ position coordinates.

It is convenient to write the above matrix expression as

$$\hat{x} = P^{-1}x \qquad (4.15)$$

where $\hat{x}$ and x are column vectors of n position coordinates $\hat{x}_i$ and the n x components of the statistical segments x_i. If Eq. (4.15) is solved for x, the result is

$$x = P\hat{x} \qquad (4.16)$$

For a linear chain as shown in Figure 7, with beads and statistical segments numbered from one end to the other, it is easy to determine the matrix P by inspection. We note that

$$\begin{aligned}
x_1 &= \hat{x}_2 - \hat{x}_1 \\
x_2 &= \hat{x}_3 - \hat{x}_2 \\
&\;\vdots \\
x_n &= \hat{x}_{n+1} - \hat{x}_n
\end{aligned} \qquad (4.17)$$

If x_1 in the first equation above is replaced by $-(x_2 + x_3 + x_4 + \cdots + x_{n+1})$, the matrix P can be written directly:

$$P = \begin{bmatrix}
2 & 1 & 1 & 1 & 1 & \cdots & \cdots & 1 \\
0 & -1 & 1 & 0 & 0 & \cdots & \cdots & 0 \\
0 & 0 & -1 & 1 & 0 & \cdots & \cdots & 0 \\
0 & 0 & 0 & -1 & 1 & \cdots & \cdots & 0 \\
\vdots & \vdots & \vdots & \vdots & \vdots & \vdots & \vdots & \vdots \\
0 & 0 & 0 & 0 & 0 & \cdots & -1 & 1
\end{bmatrix} \qquad (4.18)$$

The fact that the square matrices in Eqs. (4.14) and (4.18) are reciprocals is easily demonstrated by considering a chain of six masses and five statistical segments. The demonstration is as follows:

$$\frac{1}{6}\begin{bmatrix}
2 & 1 & 1 & 1 & 1 \\
-1 & 1 & 0 & 0 & 0 \\
0 & -1 & 1 & 0 & 0 \\
0 & 0 & -1 & 1 & 0 \\
0 & 0 & 0 & -1 & 1
\end{bmatrix}\begin{bmatrix}
1 & -4 & -3 & -2 & -1 \\
1 & 2 & -3 & -2 & -1 \\
1 & 2 & 3 & -2 & -1 \\
1 & 2 & 3 & 4 & -1 \\
1 & 2 & 3 & 4 & 5
\end{bmatrix} = \begin{bmatrix}
1 & 0 & 0 & 0 & 0 \\
0 & 1 & 0 & 0 & 0 \\
0 & 0 & 1 & 0 & 0 \\
0 & 0 & 0 & 1 & 0 \\
0 & 0 & 0 & 0 & 1
\end{bmatrix}$$

$$(4.19)$$

Now that we have available the linear transformation between the position coordinates of the masses and the coordinates of the statistical segments, we are in a position to develop an expression for the radius of gyration. We write

$$S_x{}^2 = (n + 1)^{-1} \sum \hat{x}_i{}^2 \qquad (i = 1 \text{ to } n + 1) \qquad (4.20)$$

but by using Eq. (4.6) to eliminate $\hat{x}_1$, the result is

$$\sum \hat{x}_i{}^2 = (\hat{x}_2 + \hat{x}_3 + \cdots + \hat{x}_{n+1})^2 + \hat{x}_2{}^2 + \hat{x}_3{}^2 + \hat{x}_4{}^2 + \cdots + \hat{x}_{n+1}^2 \quad (4.21)$$

Equation (4.21) indicates that $\sum \hat{x}_i$ will consist of terms such as $2\hat{x}_j{}^2$ and $2\hat{x}_i\hat{x}_j$ $(i = j)$ where every combination of i and j is represented. Equation (4.21) can thus be written in the convenient matrix form

$$\sum \hat{x}_i{}^2 = \hat{x}^T M \hat{x} \qquad (4.22)$$

where $\hat{x}^T$ is the transpose of $\hat{x}$ and M is a square matrix of order n with diagonal elements equal to 2 and all off-diagonal elements equal to 1. The radius of gyration can thus be written as

$$S_x = (n + 1)^{-1}\hat{x}^T M \hat{x} \qquad (4.23)$$

or, considering the linear transformations of Eq. (4.15),

$$S_x{}^2 = (n + 1)^{-1}x^T(P^{-1})^T M P^{-1} x \qquad (4.24)$$

For the sake of convenience the matrix product $(P^{-1})^T M P^{-1}$ is written as another matrix $(n + 1)^{-1}F$, with the result being given in the compact form

$$S_x{}^2 = (n + 1)^{-2}x^T F x \qquad (4.25)$$

The factor $(n + 1)^{-1}$ in the definition of matrix F is included so that F can be written in the simplest possible form.

It is not immediately obvious, but it is relatively easy to determine F from P and M. The general expression for the elements of the matrix, F_{ij}, is

$$
\begin{aligned}
F_{ij} &= (n + 1 - i)j \qquad \text{if } i \geq j \\
&= (n + 1 - j)i \qquad \text{if } i \leq j
\end{aligned}
\qquad (4.26)
$$

In complete detail, $\mathbf{F}$ is written

$$
\mathbf{F} = \begin{bmatrix}
-n & (n-1) & (n-2) & (n-3) & \cdots & 2 & 1 \\
(n-1) & 2(n-1) & 2(n-2) & 2(n-3) & \cdots & 4 & 2 \\
(n-2) & 2(n-3) & 3(n-2) & 3(n-3) & \cdots & 6 & 3 \\
(n-3) & 2(n-3) & 3(n-3) & 4(n-3) & \cdots & 8 & 4 \\
\vdots & \vdots & \vdots & \vdots & \vdots & \vdots & \vdots \\
3 & 6 & 9 & 12 & \cdots & 2(n-2) & (n-2) \\
2 & 4 & 6 & 8 & \cdots & 2(n-1) & (n-1) \\
1 & 2 & 3 & 4 & \cdots & (n-1) & n
\end{bmatrix}
$$

$$(4.27)$$

Equation (4.25) gives S_x^2 as the sum of squares and cross terms of the x_i. Although we know the statistical behavior of the x_i, the presence of the cross terms makes it difficult to obtain information on the statistical properties of S_x^2. To solve this problem, we transform to a coordinate system in which the expression for S_x^2 is the sum of squares of the new coordinates. We will assume that it is possible to find an orthogonal transformation matrix $\mathbf{B}$, such that

$$\mathbf{x} = \mathbf{B}\boldsymbol{\mu}$$
$$\boldsymbol{\mu} = \mathbf{B}^{-1}\mathbf{x}$$

$$(4.28)$$

When the transformation of coordinates is introduced into Eq. (4.25), the result is

$$(n+1)^2 S_x^2 = (\mathbf{B}\boldsymbol{\mu})^T \mathbf{F} \mathbf{B}\boldsymbol{\mu} = \boldsymbol{\mu}^T \mathbf{B}^T \mathbf{F} \mathbf{B}\boldsymbol{\mu}$$

$$(4.29)$$

Since $\mathbf{F}$ is a symmetric matrix, it is always possible to find a matrix $\mathbf{B}$ that will effect the transformation to a diagonal matrix $\boldsymbol{\Phi}$. The result is thus

$$(n+1)^2 S_x^2 = \boldsymbol{\mu}^T \boldsymbol{\Phi} \boldsymbol{\mu} = \Phi_1 \mu_1^2 + \Phi_2 \mu_2^2 + \Phi_3 \mu_3^2 + \Phi_4 \mu_4^2 + \Phi_5 \mu_5^2$$
$$+ \cdots + \Phi_n \mu_n^2$$

$$(4.30)$$

At this point, the problem has been simplified by transforming to a set of independent variables in which the x component of the radius of gyration is written as the sum of squares of the independent variables. Before we can proceed with the task of determining the statistical properties of S_x^2, we must learn something about the statistical properties of the μ_i. This turns out to be trivial, however, since the transformation matrix $\mathbf{B}$ is guaranteed to be orthogonal, that is,

$$\sum B_{ij}B_{jk} = 1 \quad \text{if } i = k \quad (j = 1 \text{ to } n)$$
$$= 0 \quad \text{if } i \neq k \tag{4.31}$$

It can be shown from probability theory that as long as the above relation holds, the distribution of the μ_i is the same Gaussian distribution that applies to the x_i. The implications of this equivalence are extraordinarily important. Taking the mean of both sides of Eq. (4.30) gives

$$(n+1)^2 \langle S_x^2 \rangle = \Phi_i \langle \mu_1^2 \rangle + \Phi_2 \langle \mu_2^2 \rangle + \Phi_3 \langle \mu_3^2 \rangle + \cdots \tag{4.32}$$

but since all of the $\langle \mu_i^2 \rangle$ are equal to $\langle x^2 \rangle$, the equation becomes

$$(n+1)^2 \langle S_x^2 \rangle = [\Phi_1 + \Phi_2 + \Phi_3 + \cdots]\langle x^2 \rangle \tag{4.33}$$

Theory of matrices shows that the sum of eigenvalues in the square brackets in the above equation is equal to the trace (that is, the sum of the diagonal terms) of matrix $\mathbf{F}$, which is

$$\text{tr } \mathbf{F} = n + 2(n-1) + 3(n-2) + 4(n-3) + \cdots + n \tag{4.34}$$

which can be evaluated in closed form to give

$$\text{tr } \mathbf{F} = \tfrac{1}{2}n(n+1)^2 - \tfrac{1}{6}n(n+1)(2n+1) \tag{4.35}$$

The expression for the mean square of the x component of the radius of gyration thus becomes

$$\langle S_x^2 \rangle = [\tfrac{1}{2}n - n(2n+1)/6(n+1)]\langle x^2 \rangle$$
$$= \left[\frac{n+2}{n+1}\right] \frac{n}{6} \langle x^2 \rangle \tag{4.36}$$

Expanding the term in brackets in a Taylor series in $(1/n)$ gives the following expression for $\langle S_x^2 \rangle$:

$$\langle S_x^2 \rangle = \tfrac{1}{6}[1 + n^{-1} - n^{-2} + n^{-3} - n^{-4} + \cdots]n\langle x^2 \rangle \tag{4.37}$$

Since $\langle X^2 \rangle = n\langle x^2 \rangle$, it follows that for sufficiently large n

$$\langle S_x^2 \rangle = \tfrac{1}{6}\langle X^2 \rangle \tag{4.38}$$

and

$$\langle S^2 \rangle = \tfrac{1}{6}\langle L^2 \rangle \tag{4.39}$$

4.2. The Eigenvalue Problem: Linear Chains

Up to this point we have considered only linear chains. The development has, however, been completely general for this limited class of polymer molecules. It should be clear, however, that the forms of matrices $\mathbf{F}$ and $\mathbf{P}$ are determined by the fact that they describe linear chains. Analogous matrices could be derived for chains with any kind of branching. What is required to make the statistical treatment of random-flight chains completely general is a simple prescription for finding $\mathbf{F}$ for any sort of assembly of branched molecules. Such a prescription has been worked out[15] and will be given later in this chapter. The important point is that it is the eigenvalues of $\mathbf{F}$ that determine the relationship between $\langle S^2 \rangle$ and $\langle X^2 \rangle$ (and their orthogonal components). It will also be shown that it is the same eigenvalues that determine the distribution function of S, and indeed even much of the dynamic behavior of random-flight chains in solution and in the bulk.

The diagonalization of $\mathbf{F}$ is, on the surface, a formidable problem. Indeed the first published eigenvalues[16] were correct only in the limit as $n \rightarrow \infty$. The limiting eigenvalues are, in fact, applicable to most problems for which the random-flight chain is intended to model the behavior of high-molecular weight polymer molecules. The appropriate place to start the analysis is, however, to consider the correct, closed-form values of Φ_i. It was recognized at the time that $\mathbf{F}$ was introduced[16] that its inverse had the following simple form:

$$\mathbf{F}^{-1} = (n+1)^{-1} \begin{bmatrix} 2 & -1 & 0 & 0 & 0 & \cdots & \cdot & \cdot & \cdot \\ -1 & 2 & -1 & 0 & 0 & \cdots & \cdot & \cdot & \cdot \\ 0 & -1 & 2 & -1 & 0 & \cdots & \cdot & \cdot & \cdot \\ 0 & 0 & -1 & 2 & -1 & \cdots & \cdot & \cdot & \cdot \\ \vdots & \vdots & \vdots & \vdots & \vdots & \vdots & \vdots & \vdots & \vdots \\ 0 & 0 & 0 & \cdot & \cdot & \cdots & -1 & 2 & -1 \\ 0 & 0 & 0 & \cdot & \cdot & \cdots & 0 & -1 & 2 \end{bmatrix} \quad (4.40)$$

It will be demonstrated later in this chapter that the above tridiagonal matrix appears in the study of the dynamics of random-flight chains. It was introduced by Rouse[17] in the solution of what would seem to be an entirely different problem. In the course of the solution of this problem, he gave its eigenvalues in closed form. We will look more carefully into this aspect of polymer theory later in this chapter, but for now we will just assert that since the eigenvalues of the Rouse tridiagonal matrix are known, we also know the eigenvalues of $\mathbf{F}$, since the eigenvalues of matrices which are reciprocals are the reciprocals of each other. We can thus write

$$\Phi_j = [(n + 1)/4]\{\sin[j\pi/2(n + 1)]\}^{-2} \tag{4.41}$$

Although the eigenvalues given above are correct, it is useful to consider the limiting form taken by Eq. (4.41) as n becomes very large. For that special case

$$\Phi_j = n^3/\pi^2 j^2 \tag{4.42}$$

Equation (4.42) applies, of course, only for $n \gg j$. It is useful, however, to test this relationship for random flight chains of finite length to see when it might be applicable. For example, if $n = 100$, the eigenvalues given by Eq. (4.42) are correct to within a fraction of a percent up to $j = 25$. If we use these limiting eigenvalues in Eq. (4.32) and let $n = (n + 1)$, we get the following interesting expression for S_x^2:

$$S_x^2 = (n/\pi^2)[\mu_1^2 + (1/4)\mu_2^2 + (1/9)\mu_3^2 + (1/16)\mu_4^2 \\ + (1/25)\mu_5^2 + \cdots] \tag{4.43}$$

Equation (4.43) illustrates an important feature of this theoretical analysis. If we number the normal coordinates the same way we number the eigenvalues, the importance of the normal coordinate in determining the statistics of S_x (and S) decreases rapidly with increasing order. This can be clearly shown by using Eq. (4.43) to determine $\langle S_x^2 \rangle$. We find that

$$\langle S_x^2 \rangle = (n/\pi^2)[1 + (1/4) + (1/9) + (1/16) + (1/25) \\ + (1/36) + \cdots]\langle \mu^2 \rangle \tag{4.44}$$

If the chain is considered to be infinite, the sum in the square brackets can be written in closed form—it is equal to $(\pi^2/6)$. For the infinite chain, then, we get

$$\langle S_x^2 \rangle = (1/6)n\langle \mu^2 \rangle = (1/6)n\langle x^2 \rangle \tag{4.45}$$

which, of course, is the correct answer in the limit as the chain length approaches infinity. If, however, we truncate the sum in the square brackets of Eq. (4.44) after the fifth term, the computed value of the $\langle S_x^2 \rangle$ is 2.5% low. One might interpret this to mean that 97.5% of the statistics of the radius of gyration is determined by the first five normal coordinates. In succeeding sections we will see that the normal coordinates that enter into chain statistics in a very natural way are identical to the normal coordinates that are used to describe the dynamic behavior of random-flight chains.

4.3. The Distribution of the Radius of Gyration: One Approach

It was shown in the previous section that the mean square of the radius of gyration can be determined from the eigenvalues of $\mathbf{F}$ and the mean-square value of the x_i. Indeed it was asserted that this was true for branched chains as well as linear chains. Although it is true that much of the physics of polymer molecules is dominated by the mean-square values of S and/or S_x, it is also true that knowledge of mean values is imperfect knowledge at best, and that it is important to know the complete statistical behavior through the distribution function. It is not surprising, therefore, that there has been a continuing interest in determining the distribution functions associated with the radii of gyration. The research in this area is well documented,[3] and only one approach will be described here. This theoretical approach was chosen, however, because it demonstrates a unity between chain statistics and chain dynamics which is not always appreciated.

Consider a random variable, X, which is the sum of independent random variables, x_i, that is,

$$X = x_1 + x_2 + x_3 + \cdots \tag{4.46}$$

It was shown in Section 2.3.5 that the characteristic function of the distribution of x, $f(X)$, is the product of the characteristic functions of the distributions $f_i(x_i)$. That is

$$\bar{f}(\omega) = \bar{f}_1(\omega)\bar{f}_2(\omega)\bar{f}_3(\omega)\bar{f}_4(\omega) \ldots \tag{4.47}$$

If we look again at Eq. (4.30) we can see how Eqs. (2.10) and (2.11) can be used to obtain the distribution function of S_x, since

$$S_x{}^2 = (n + 1)^{-2} \sum \Phi_j \mu_j{}^2 \tag{4.48}$$

Equation (4.48) shows that the square of the x component of the radius of gyration is the sum of random variables which are themselves the squares of the normal coordinates multiplied by the associated eigenvalue and divided by $(n + 1)^2$. Since the distribution functions of the μ_j are known, it is easy to determine the distribution functions for the random variable Γ_j, where

$$\Gamma_j = \Phi_j \mu_j{}^2/(n + 1)^2 \tag{4.49}$$

Indeed, if W_j is the distribution function of Γ_j, we can write

$$W_j(\Gamma_j) = [\beta(n + 1)/(\pi\Phi_j)^{1/2}] \exp[-\beta^2(n + 1)^2\Gamma_j/\Phi_j] \tag{4.50}$$

and

$$\overline{W}_j(\omega) = [1 + (i\Phi_j\omega)/(n+1)]^{1/2} \qquad (4.51)$$

It is now possible to put down the expression for $\overline{W}(\omega)$, the characteristic function for the distribution of $S_x{}^2$:

$$\overline{W}_{S_x{}^2}(\omega) = \frac{1}{\prod [1 + (i\Phi_j\omega)/(n+1)]^{1/2}} \qquad (4.52)$$

The index j in the continued product runs from 1 to n. And, of course, the distribution of $S_x{}^2$ can then be obtained by inverting its Fourier transform (i.e., characteristic function):

$$W_{S_x{}^2}(S_x{}^2) = \frac{1}{2\pi} \int_{-\infty}^{+\infty} \frac{\exp[-in(n+1)\beta^2 S_x{}^2 \omega]}{\prod [1 + (i\Phi_j\omega)/(n+1)]^{1/2}} \, d\omega \qquad (4.53)$$

where the above distribution function applies to the dimensionless random variable $n(n+1)\beta^2 S_x{}^2$.

This same approach can be used to find the distribution function for S^2 itself. Since $S^2 = S_x{}^2 + S_y{}^2 + S_z{}^2$, the characteristic function of the distribution of S^2 is just the cube of that given in Eq. (4.52). Equations (4.52) and (4.53) were derived by various methods and approximate and asymptotic solutions given long before complete solutions were found.[16,18,19]

It is important to note that this approach to the distribution functions for the radii of gyration is perfectly general and would apply to any kind of branching and to chains of any length. The only mathematical difficulties involve diagonalization of **F** to find the eigenvalues Φ_i, and evaluation of integrals of the type shown in Eq. (4.53). These can, however, be formidable problems indeed. The first complete solution[20] was for linear chains in the limit as n approaches infinity, that is, the eigenvalues used in Eqs. (4.52) and (4.53) were the limiting ones given in Eq. (4.42). For this special case, Eq. (4.53) becomes

$$W_{S_x{}^2}(S_x{}^2) = \frac{1}{2\pi} \int_{-\infty}^{+\infty} \frac{\exp[-in(n+1)\beta^2 S_x{}^2 \omega]}{\prod [1 + (in^2\omega)/\pi^2 j^2]^{1/2}} \, d\omega \qquad (4.54)$$

Likewise, for the polar radius of gyration

$$W_{S^2}(S^2) = \frac{1}{2\pi} \int_{-\infty}^{+\infty} \frac{\exp[-in(n+1)\beta^2 S^2 \omega]}{\prod [1 + (in^2\omega)/\pi^2 j^2]^{3/2}} \, d\omega \qquad (4.55)$$

The integrals in Eqs. (4.54) and (4.55) cannot be evaluated in closed form. They were, however, determined numerically using several techniques.

In addition, the asymptotic forms for both large and small radii were explored in some detail. Finally, the distribution function for the two-dimensional radius of gyration can be written in closed form. In summary, it is fair to say that the distribution functions for the one-, two-, and three-dimensional radius of gyration are known in considerable detail for sufficiently long, linear random-flight chains. Results of the computations are shown in considerable detail in the original publication.[20]

4.4. The Distribution of the Radius of Gyration: A Second Approach

When we consider the problem of chain statistics in three dimensions, we must remember that there are only $3n$ degrees of freedom with which the shape of the chain can be defined. These $3n$ degrees of freedom are the n values of x_i, the n values of the y_i, and the n values of the z_i—defining these $3n$ numbers completely defines the shape of the random-flight chain. If we examine chain statistics from this point of view, and if we assume that there is no energy of interaction between any of the masses of a given chain (or, for that matter, between masses of different chains), it is clear that the distribution function characterizing an assembly of these molecules, W_{xyz}, is the product of the distribution functions of the $3n$ independent random variables that define the shape of the chain. That is,

$$W_{xyz} = (\sqrt{n}\,\beta/\sqrt{\pi})^{3n} \exp[-n\beta^2(\mathbf{x}^T\mathbf{x} + \mathbf{y}^T\mathbf{y} + \mathbf{z}^T\mathbf{z})] \qquad (4.56)$$

Since it is the radius of gyration which is of interest, and it is defined in terms of the position of the masses relative to the center of mass of the molecule, it is useful to transform Eq. (4.56) to the distribution of the position variables $\hat{\mathbf{x}}$, $\hat{\mathbf{y}}$, and $\hat{\mathbf{z}}$. This transformation is easily accomplished using Eq. (4.16). The result is

$$W_{xyz} = C \exp[-n\beta^2(\hat{\mathbf{x}}^T\mathbf{P}^T\mathbf{P}\hat{\mathbf{x}} + \hat{\mathbf{y}}^T\mathbf{P}^T\mathbf{P}\hat{\mathbf{y}} + \hat{\mathbf{z}}^T\mathbf{P}^T\mathbf{P}\hat{\mathbf{z}})] \qquad (4.57)$$

It should be remembered, of course, that only $3n$ out of the $3(n+1)$ position coordinates appear in Eq. (4.57). The normalization constant C will not be considered for the moment.

The distribution function for S^2 can be obtained from Eq. (4.57) by multiplying W_{xyz} by a Dirac delta function centered on S^2 and then integrating over all of the position coordinates $\hat{x}_i$, $\hat{y}_i$, and $\hat{z}_i$. (This procedure eliminates all conformations from the distribution except those associated with a given radius of gyration.) The form of the delta function most convenient for this purpose is that given in a Fourier transform representation. Since S^2 is expressed in terms of the position coordinates in Eq. (4.57), this form of the delta function is written as

$$\int_{-\infty}^{+\infty} \exp\left\{2\pi i \omega\left[\frac{\hat{\mathbf{x}}^T \mathbf{M}\hat{\mathbf{x}} + \hat{\mathbf{y}}^T \mathbf{M}\hat{\mathbf{y}} + \hat{\mathbf{z}}^T \mathbf{M}\hat{\mathbf{z}}}{n+1} - S^2\right]\right\} d\omega \qquad (4.58)$$

where $\mathbf{M}$ was defined in Eq. (4.22). The above form of the delta function is displayed in a common "shorthand" notation. In point of fact, the function to be operated upon is first multiplied by the integrand in the above expression, and then the integration is performed. When this operation is carried out with the distribution in Eq. (4.57), the result is

$$W_{S^2}(S^2) = C\int_{-\infty}^{+\infty}\int \cdots \int_{-\infty}^{+\infty} \exp[-n\beta^2(\hat{\mathbf{x}}^T \mathbf{P}^T \mathbf{P}\hat{\mathbf{x}} + \hat{\mathbf{y}}^T \mathbf{P}^T \mathbf{P}\hat{\mathbf{y}} + \hat{\mathbf{z}}^T \mathbf{P}^T \mathbf{P}\hat{\mathbf{z}})]$$

$$\exp\left\{2\pi i \omega\left[\frac{\hat{\mathbf{x}}^T \mathbf{M}\hat{\mathbf{x}} + \cdots}{n+1} - S^2\right]\right\} d\omega \, d\hat{\mathbf{x}} \, d\hat{\mathbf{y}} \, d\hat{\mathbf{z}} \qquad (4.59)$$

It is convenient to rewrite Eq. (4.59) in a somewhat more compact form by defining a new matrix $\mathbf{D}$, where

$$\mathbf{D} = \mathbf{P}^T \mathbf{P} - i\omega \mathbf{M} \qquad (4.60)$$

Using $\mathbf{D}$ in Eq. (4.59) gives

$$W_{S^2}(S^2) = C\int_{-\infty}^{+\infty}\int \cdots \int \exp[-n\beta^2(\hat{\mathbf{x}}^T \mathbf{D}\hat{\mathbf{x}} + \hat{\mathbf{y}}^T \mathbf{D}\hat{\mathbf{y}} + \hat{\mathbf{z}}^T \mathbf{D}\hat{\mathbf{z}})]$$

$$\exp[-in(n+1)\beta^2 S^2 \omega] \, d\omega \, d\hat{\mathbf{x}} \, d\hat{\mathbf{y}} \, d\hat{\mathbf{z}} \qquad (4.61)$$

As complicated as the integral in Eq. (4.61) appears, it is possible to express the integration with respect to the position coordinates in simple, closed form. To carry out the integration, we switch to a new set of coordinates—normal coordinates—through an orthogonal transformation. Indeed, we will see that the same orthogonal matrix, $\mathbf{B}$, works for this transformation as well as for the transformation that reduces the expression for S^2 to the sum-of-squares of normal coordinates. One should note, however, that in the previous application of coordinate transformation we were dealing with the x, y, and z components of the statistical segment vectors, where in this application we are dealing with the x, y, and z components of the position vectors of the beads relative to the center of mass of the chain [but using only $3n$ out of the $3(n + 1)$ total coordinates]. To illustrate how this works, we begin by introducing the following transformation of x components of the position coordinates:

$$\hat{x} = \mathbf{B}\boldsymbol{\Gamma} \tag{4.62}$$

With this transformation, the quadratic form in $\hat{x}^T \mathbf{D}\hat{x}$ becomes

$$\hat{x}^T \mathbf{D}\hat{x} = \boldsymbol{\Gamma}^T \mathbf{B}^T \mathbf{P}^T [\mathbf{M} - i\omega(n + 1)^{-1}\mathbf{F}]\mathbf{P}\mathbf{B}\boldsymbol{\Gamma} \tag{4.63}$$

Furthermore, we can take advantage of the fact that

$$\mathbf{M} = \boldsymbol{\delta} + \mathbf{1} \tag{4.64}$$

where $\boldsymbol{\delta}$ is the matrix with diagonal elements equal to unity and all of the other elements equal to zero, and $\mathbf{1}$ is a matrix with all of its elements equal to unity. Furthermore, since $\mathbf{B}$ is an orthogonal matrix, it can be shown that

$$\mathbf{B}^T \mathbf{1}\mathbf{B} = \mathbf{0} \tag{4.65}$$

where $\mathbf{0}$ is a matrix with all of its elements equal to zero. By capitalizing on these useful properties of $\mathbf{M}$, Eq. (4.63) can be transformed into

$$\hat{x}^T \mathbf{D}\hat{x} = \boldsymbol{\Gamma}^T \mathbf{B}^T \mathbf{P}^T [\boldsymbol{\delta} - i\omega(n + 1)^{-1}\mathbf{F}]\mathbf{P}\mathbf{B}\boldsymbol{\Gamma} \tag{4.66}$$

If we keep in mind that $\mathbf{B}^T \mathbf{B} = \boldsymbol{\delta}$, and from Eq. (4.30) and Eq. (4.31) that $\mathbf{B}^T \mathbf{F}\mathbf{B} = \boldsymbol{\Phi}$, Eq. (4.66) gives

$$\hat{x}^T \mathbf{D}\hat{x} = \boldsymbol{\Gamma}^T \boldsymbol{\mu} \boldsymbol{\Gamma} \tag{4.67}$$

where $\boldsymbol{\mu}$ is a diagonal matrix with elements given by

$$\mu_j = 1 - i\omega(n + 1)^{-1}\omega_j \tag{4.68}$$

This orthogonal transformation with matrix $\mathbf{B}$ works the same way, of course, with the y and z components of the position coordinates. So if this operation is performed on all of the position coordinates in the three orthogonal directions, the net effect is the transformation of the quadratic form in the first exponential of Eq. (4.61) to the sum of squares of terms such as Γ_j. The normal coordinates then become the variables of integration. Since the exponential now contains the sum of squares of the variable of integration, it can be factored into the product of Gaussian functions, each of which can be integrated as follows:

$$\int_{-\infty}^{+\infty} \exp\{-n\beta^2[1 - i\omega(n+1)^{-1}\omega j]\Gamma_j^2\}\, d\Gamma$$

$$= \frac{\sqrt{\pi}}{\sqrt{n}\,\beta[1 - i\omega(n+1)^{-1}\omega_j]^{1/2}} \tag{4.69}$$

After the $3n$ integrations over the position coordinates have been performed, the final result is

$$W_{S^2}(S^2) = C \int_{-\infty}^{+\infty} \frac{\exp[-in(n+1)\beta^2 S^2\omega]}{\prod [1 + (i\Phi_j\omega)/(n+1)]^{3/2}}\, d\omega \tag{4.70}$$

Equation (4.70) is, of course, the same as one would get by extending equation (4.53) to three dimensions, and we can identify the normalization constant as $(1/2\pi)$. One would be justified in questioning why it is of interest to display both methods of arriving at the same expression for the characteristic function of the distribution of the square of the radius of gyration. Looking at this second method does, however, illustrate the connection between the normal coordinate used to express the radius of gyration as the sum of squares and how these same normal coordinates are related to the effect of segment–segment interactions on the dimensions of polymer molecules when such interactions are present. This, however, cannot be appreciated until we consider the excluded volume problem, which will be taken up next.

5. EXCLUDED VOLUME

Up to this point, the development has been based on the assumption that there is no energy of interaction between the beads that represent the mass of the polymer molecule in the random-flight model. One of the implications of this assumption is that the centers of mass of two or more beads can be located at the same point at the same time. If we allow ourselves to be preoccupied with the pictorial aspect of the spring-bead model, and assume that each bead represents a region in space fully occupied with polymer segments, this assumption seems totally unrealistic—clearly no two beads could occupy the same volume at the same time. One must remember, however, that the mass of the polymer chain represented by a given bead *is itself* a length of chain which, for flexible polymer molecules,

repeat units of bead p   repeat units of bead q

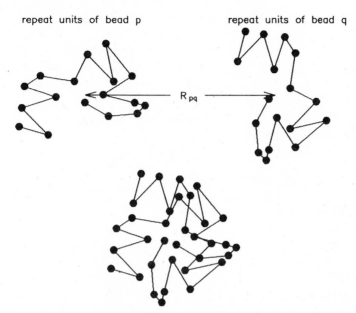

R_{pq}

FIGURE 8. Repeat units that make up beads p and q when their centers of mass are a distance R_{pq} apart (above), and when their centers of mass are located at the same point in space (below).

is coiled around in space. So what is pictorially represented as a bead is, in terms of the model, a region in space rather sparsely occupied by the number of repeat units that constitute one statistical segment. Clearly the centers of mass of two such sparsely populated beads, either from the same molecule or from different molecules, could be located at the same point at the same time. This notion is illustrated in Figure 8, which represents two beads p and q a distance $R_{pq} \neq 0$ apart and when $R_{pq} = 0$.

The next logical extension of the model is the introduction of a potential of mean force acting between the centers of mass of any pair of beads. Since bringing their centers of mass into close proximity results in an increase in the concentration of repeat units in the neighborhood of the pair of beads, there will be changes in both the energy and entropy of the system. The potential of mean force between two beads, F_{pq}, will be a function of the distance between their centers of mass and will have energetic and entropic components.

The term *entropy* is used here in a special way. Entropy is, of course, a concept originally defined for macroscopic systems, and we are dealing here with systems of molecular dimensions—namely, pairs of subchains associated with pairs of beads. So to be consistent with the more common terminology, one should say that both the energy of the system and the number of ways the subchains can be accommodated in the system depend upon how the statistical segments are fixed with respect to length and orientation. If, however, we define Ω as the number of ways the chains can be accommodated in the system, we can use the Boltzmann equation to *define* a quantity $S = k \ln \Omega$ that we will call the entropy. The *weighted* Boltzmann factor can then be written

$$\Omega e^{-E/kT} \tag{5.1}$$

$$e^{-(E-TS)/kT} \tag{5.2}$$

$$e^{-F/kT} \tag{5.3}$$

We can thus combine the energy and combinatorial effects and say that the potential of mean force acting between any two beads p and q is a *free energy*.

5.1. The Perturbation Method: Linear and Branched Chains

In the analysis to follow, we will be considering polymers at infinite dilution. Only intramolecular bead–bead interactions need to be considered. The total free energy of bead–bead interaction is thus the sum of terms F_{pq} that apply to the bead pairs in an individual chain molecule. We can thus write

$$F = \sum_{p=1}^{n} \sum_{q=p+1}^{n+1} F_{pq} \tag{5.4}$$

Each F_{pq} is a function only of R_{pq}. It is important to note, that although the indices p and q run from 1 to $n + 1$, the potential functions must always be expressed in terms of n *independent* position vectors.

Given the potential of mean force F, the distribution function for the radius of gyration is modified by including a Boltzmann factor. Equation (4.57) is replaced by

$$W_{S^2}(S^2) = C \exp[-n\beta^2(\hat{\mathbf{x}}^T \mathbf{P}^T \mathbf{P}\hat{\mathbf{x}} + \hat{\mathbf{y}}^T \mathbf{P}^T \mathbf{P}\hat{\mathbf{y}} + \hat{\mathbf{z}}^T \mathbf{P}^T \mathbf{P}\hat{\mathbf{z}})] \exp(-F/kT) \tag{5.5}$$

It is convenient to write each of the $\exp(-F_{pq}/kT)$ terms in a special form before operating on Eq. (5.2) any further. We define a set of functions f_{pq} by the relationship

$$f_{pq} = \exp(-F_{pq}/kT) - 1 \tag{5.6}$$

With this definition

$$\exp(-F/kT) = (1 + f_{pq})$$
$$= 1 + \Sigma f_{pq} + \cdots \qquad (5.7)$$

There are $n(n + 1)/2$ terms of the type f_{pq} in Eq. (5.7), each one associated with the interaction of a pair of beads. There will be higher-order terms which consist of the products of the various f_{pq}, but these will be neglected in this analysis.

Equation (5.7) is substituted into Eq. (5.5). This expression is then multiplied by a delta function of the form shown in Eq. (2.58) to give the distribution of S^2 in the presence of the intramolecular interactions in terms of the following integral:

$$W_{S^2}(S^2) = C \int_{-\infty}^{+\infty} \int \cdots \int \exp[-n\beta^2(\hat{\mathbf{x}}^T\mathbf{D}\hat{\mathbf{x}} + \hat{\mathbf{y}}^T\mathbf{D}\hat{\mathbf{y}} + \hat{\mathbf{z}}^T\mathbf{D}\hat{\mathbf{z}})]$$
$$\exp[-in(n + 1)\beta^2 S^2\omega][1 + \Sigma f_{pq} + \cdots] \, d\omega \, d\hat{\mathbf{x}} \, d\hat{\mathbf{y}} \, d\hat{\mathbf{z}} \qquad (5.8)$$

Integration of Eq. (5.8) is beyond the scope of this chapter.[3] Zimm, Stockmayer, and Fixman[21] and Fixman[22] were the first to solve the problem and gave the results in the form

$$\alpha_S^2 = 1 + \frac{134}{105} Z - 2.082Z^2 + \cdots \qquad (5.9)$$

where

$$\alpha_S^2 = \langle S^2 \rangle / \langle S^2 \rangle_0 \qquad (5.10)$$

and

$\langle S^2 \rangle$ = mean-square radius of gyration in the presence of the bead–bead interactions

$\langle S^2 \rangle_0$ = mean-square radius of gyration in the absence of the bead–bead interactions, i.e., the unperturbed mean-square radius of gyration

The argument Z of the power series in Eq. (5.9) is given by

$$Z = n^{1/2}(n/4\pi\langle S^2 \rangle_0)^{3/2}b \qquad (5.11)$$

where

$$b = -\int_V f_{pq} \, dV \qquad (5.12)$$

the quantity b is called the binary cluster integral: the integration is over all space enclosing the molecule, but the integrand f_{pq} is zero for values of R_{pq} much greater than the root-mean-square length of one statistical segment.

An implication in expressing the result of the analysis in the form of Eq. (5.9) is that, when Z is sufficiently small, $\alpha_S{}^2$ can be expanded as a Taylor series in Z. Gordon[23] and Eichinger[24] and their associates have demonstrated that this is not as straightforward a notion as one might expect, and that convergence of the series can be guaranteed only if the potential of mean force $F(R_{pq})$ conforms to certain rather strict mathematical requirements. For our purposes, we will assume that Eq. (5.9) gives the expansion factor correctly, at least to the first power in Z. The more important aspect of the problem is, however, that the Taylor series representation of Eq. (5.9), if it actually exists, converges so slowly that it is useless as a representation of $\alpha_S{}^2$ except over that very narrow range of physical conditions for which $Z \leq 0.1$.

The same sort of analysis can be carried out for the distribution of end-to-end distances with the result

$$\alpha_L{}^2 = 1 + \tfrac{4}{3}Z - \cdots \tag{5.13}$$

where

$$\alpha_L{}^2 = \langle L^2 \rangle / \langle L^2 \rangle_0 \tag{5.14}$$

and

$\langle L^2 \rangle$ = mean-square end-to-end distance in the presence of the bead–bead interactions

$\langle L^2 \rangle_0$ = mean-square end-to-end distance in the absence of the bead–bead interactions, i.e., the unperturbed mean-square end-to-end distance

The analyses that led to Eq. (5.9) and (5.13) were based on the eigenvalues for the infinite chain. Computations of the expansion factor for the radius of gyration have, however, since been done using the eigenvalues for finite chains.[25] Table 1 gives the values of the coefficient of Z, C_n, as a function of the number of statistical segments for values of n up to 100. The results were also extrapolated to $n = 1000$. When $n = 10$, 100, and 1000, the coefficient of Z is about 30%, 10%, and 5% lower than the limiting value of $(134/105) = 1.276$. It is clear that, except for very high molecular weights, there is a chain-length effect on the expansion factor of real polymers in addition to the effect that enters through Z. This problem has never

TABLE 1. Coefficients of Z from Perturbation Calculations as a Function of the Number of Statistical Segments

n	C_n	n	C_n
2	0.730	20	0.979
4	0.785	30	1.028
6	0.830	40	1.058
8	0.866	50	1.079
12	0.918	70	1.107
16	0.954	100	1.133
		∞	1.275[a]

[a] Fixman, Reference 22.

been addressed, since all treatments of chain expansion have incorporated eigenvalues in the limit as $n \to \infty$.

We will consider two special cases in the application of Eq. (5.9) and (5.13). Since the F_{pq} have both entropy and energy contributions, there exists the possibility that, at a given temperature, they will cancel, and F_{pq} will be equal to zero for all R_{pq}. If this is the case, it is clear that the chains will demonstrate unperturbed, random-flight statistics, as was described in Section 4. This represents the theta condition as defined by Flory. It is also possible that, even though the F_{pq} are not zero, the balance of energy and entropy give a potential with attractive and repulsive contributions such that the binary cluster integral vanishes. For that special case $Z = 0$, *and to a first approximation*, the mean-square radius of gyration is equal to its unperturbed value. It has been shown, however, that, if this is the case, the chains would not demonstrate unperturbed, random-flight statistics.[26] This second case has been referred to as the theta state and the first as the unperturbed state.[26]

In this author's view, this switch in nomenclature is inappropriate, since as will be shown shortly, the original Flory model implies that the theta condition is equivalent to $F_{pq} = 0$. Vanishing of Z without the corresponding vanishing of F_{pq} is an intriguing notion and leads to some interesting predictions. Within the general framework of the model for polymer–polymer and polymer–solvent interactions introduced by Flory, this sort of potential function is impossible and such behavior is unlikely to be demonstrated by any real polymer system. In any event, this topic is outside of the scope of this chapter.

5.2. The Flory–Huggins Model for Polymer–Solvent Interactions

So far we have considered the potential of mean force in a rather general way without considering exactly what form it might take. It would be useful, however, if $F(R_{pq})$ could be related to other parameters associated with the thermodynamics of polymer solutions. To make this connection we will review the now classical Flory–Huggins[27] theory for the solution thermodynamics of monodisperse chain molecules. Although imprecise in many aspects, the Flory–Huggins theory has a long history of application and extensive correlation with experimental data. The unique feature of the theoretical development is the determination of the entropy of mixing of an assembly of polymer chains and solvent. This is done by counting the number of ways solvent molecules and polymer segments can be accommodated on a lattice. According to the Flory–Huggins theory the free energy of mixing, ΔF_m is written

$$\Delta F_m/kT = [n_1 \ln(v_1) + n_2 \ln(v_2) + n_1 v_2 \chi] \tag{5.15}$$

where n_1 and n_2 are the number of solvent and polymer molecules and v_1 and v_2 are the volume fraction of solvent and polymer, respectively. Since the theory assumes that there is no change in volume on mixing, ΔF_m represent changes in both Gibbs and Helmholtz free energies. The term χ is a van Laar interaction parameter which is made dimensionless by dividing by kT. It is a measure of energy (and enthalpy) of mixing. Indeed, it represents the change in enthalpy when one mole of Flory–Huggins (FH) segments is mixed with an infinite volume of solvent. One FH segment is defined as that length of a polymer molecule occupying a volume equal to the volume of one solvent molecule. If r is the number of FH segments per chain, it is easy to show that

$$r = \bar{v}M/V_1 \tag{5.16}$$

where $\bar{v}$ is the specific volume of the polymer, M is its molecular weight, and V_1 is the molar volume of the solvent. Indeed, one might note that r is just the ratio of the molar volumes of polymer and solvent. Within the framework of this theory, the enthalpy of mixing of a unit volume of solution is given by

$$\Delta H/kT = n_1 v_2 \chi \tag{5.17}$$

The contribution $[n_1 \ln(v_1) + n_2 \ln(v_2)]$ to $\Delta F_m/kT$ in Eq. (5.15) represents the entropy of mixing of an assembly of random-flight chain molecules with the solvent, that is, the combinatorial effect. It is important to note, however, that Hildebrand[28] derived the same expression in a much more general way without resorting to lattice counting techniques. There is thus reason to believe that this entropy term is more general than one would expect from the assumptions introduced in the development of the Flory–Huggins theory.

There are two quite different types of limitations to the Flory–Huggins theory. The first type includes the same sort of limitations associated with any simple statistical mechanical theory of mixing of small-molecule species. It was assumed that there were no energetically preferred arrangements of the segments and solvent molecules on the lattice, i.e., energetic effects were assumed to be too small to induce segment–segment, solvent–solvent, or segment–solvent association. This assumption would, of course, apply only to the most nonpolar polymer–solvent systems. Energetic effects were also assumed to be concentration independent and could be adequately described by a single van Laar parameter χ. These limitations thus cloud the interpretation of experimental measurements in terms of the Flory–Huggins theory. They can, however, be accommodated to some extent by making the additional assumptions that χ has an entropic component (and that it thus has a component that is temperature independent) and that it is a function of concentration. The entropic contribution to χ is thought to derive from short-range ordering or disordering associated with polymer–solvent interactions. This empirical interpretation of χ leaves intact the notion that the terms $n_1 \ln(v_1) + n_2 \ln(v_2)$ are derived from the entropy of dispersing the random flight chains throughout the solution.

The second limitation has to do with the range of concentrations over which the Flory–Huggins theory can be expected to apply. The derivation of Eq. (5.15) applies *only* to solutions sufficiently concentrated that they can be considered to have a uniform density of FH segments throughout. This is to be contrasted with solutions that are sufficiently dilute that each individual polymer molecule is separated from its neighbors by a sea of pure solvent. Keeping these two limitations of the Flory–Huggins theory in mind, we will proceed with the analysis.

Taking the partial derivative of Eq. (5.15) with respect to n_1 gives the chemical potential of the solvent in the solution relative to that of pure solvent. The chemical potential of the solvent is directly related to the osmotic pressure π. The result of this exercise gives

$$\pi = -(RT/V_1)[\ln(v_1) + (1 - r^{-1})v_2 + \chi v_2{}^2] \qquad (5.18)$$

Since $v_1 = 1 - v_2$, the term $\ln(v_1)$ can be expanded in a Taylor series in v_2 and the relationship given in Eq. (5.18) becomes

$$\pi = (RT/V_1)[(v_2/r) + (\tfrac{1}{2} - \chi)v_2{}^2 + \tfrac{1}{3}v_2{}^3 \cdots] \qquad (5.19)$$

Since the assumption is also made that there is no change of volume upon mixing, one can write

$$v_2 = C\bar{v} \qquad (5.20)$$

where C is the polymer concentration in units reciprocal to those of $\bar{v}$. Given these relationships, Eq. (5.19) can be recast in the following form:

$$(\pi/C) = (RT/M)[1 + (M\bar{v}^2/V_1)(\tfrac{1}{2} - \chi)C + \tfrac{1}{3}(M\bar{v}^3/V_1)C^2 \cdots] \qquad (5.21)$$

There are a number of very interesting points to be made about Eq. (5.21). The limit of (π/C) as $C \to 0$ is (RT/M) just as it must be to satisfy the van't Hoff law. But Eq. (5.21) has some serious deficiencies. The second virial coefficient, $(M\bar{v}^2/V_1)(\tfrac{1}{2} - \chi)$ should be the initial slope in an extrapolation to infinite dilution. Unfortunately, such an extrapolation goes into a concentration regime too dilute for Eq. (5.15) to apply. On the other hand, one could interpret Eq. (5.21) as applying only at high enough concentrations for the uniform segment density model to apply. But for concentrations much greater than dilute solutions, v_2 is too large for the assumption that χ is independent of concentration to apply.[1] The original Flory–Huggins theory, while embodying the essential features of polymer solution thermodynamics, is, in principle, not capable of giving a rigorous, quantitative description of polymer solution thermodynamics. It has, however, served as a useful semiempirical relationship for correlating experimental data.

Flory and Krigbaum[29] introduced a remedy for this situation which preserved the concept of a polymer–solvent interaction parameter χ. They pointed out that a simple variation of Eq. (5.15) can be used to describe the free energy of mixing of *any* region of space in which lengths of polymer chain (either from different molecules or from the same molecule) are mixed with solvent. That simple variation is

$$\delta(\Delta F_m/kT) = [v_1 \ln(v_1) + v_1 v_2 \chi](\delta V/V_S) \qquad (5.22)$$

where δV is the element of volume under consideration and V_S is the volume of a solvent molecule or FH segment. Since $v_1 = 1 - v_2$, the term $\ln(v_1)$ can be expanded in a Taylor series in v_2 in the element of volume δV.

The result is

$$\delta(\Delta F_m/kT) = [(\chi - 1)v_2 + (\tfrac{1}{2} - \chi)v_2^2 + \cdots](\delta V/V_S) \qquad (5.23)$$

Since we will be considering elements of volume for which v_2 is small, it is justifiable to truncate the above series after the second-order term in v_2.

Equation (5.15) is intended to apply to elements of volume of molecular dimensions. The entropic contribution $v_1 \ln(v_1)$ must thus be considered as a combinatorial term accounting for the number of ways the lengths of chain and solvent molecules can be accommodated in the element of volume δV. This expression for $\delta(\Delta F_m/kT)$ can be used to calculate the potential of mean force in the perturbation treatment given in Section 5.1. Before going back to the perturbation theory we will, however, examine dilute solution thermodynamics in a little more detail.

Flory and Krigbaum[29] used Eq. (5.23) to derive an equation for the osmotic pressure which is applicable in the limit as $C \to 0$. They used a statistical mechanical treatment introduced by McMillan and Mayer[30] that gives the osmotic pressure of dilute solutions given the potential of mean force between solute molecules. Indeed the procedure is formally the same as that for determining the pressure–volume–temperature relationship of dilute, but nonideal, gasses. The key step in their development was the derivation of a potential of mean force between the centers of mass of two polymer molecules. An essential feature of the theory was the model they used for the mean distribution of FH segments about the centers of mass of the polymer molecules. They took the distribution to be Gaussian. If the segment density is ϱ, their expression is

$$\varrho = r\left(\frac{\sqrt{6}\,\beta}{\alpha\sqrt{\pi}}\right)^3 e^{-6\beta^2 s^2/\alpha^2} \qquad (5.24)$$

Although the true distribution is not quite Gaussian, Eq. (5.24) is a reasonable approximation. When the parameter α is equal to unity the density given in Eq. (5.24) gives the correct mean-square radius gyration of the assembly of unperturbed polymer chains. With this expression for ϱ, α is thus the measure of the expansion of the polymer chains beyond their unperturbed dimensions. In these early developments there was no distinction made between expansion in mean-square radius of gyration and mean-square end-to-end distance, so the expansion factor as defined in Eq. (5.24) is not given a subscript.

The potential of mean force acting between a pair of molecules p and q is obtained by integrating $\delta(\Delta F_M/kT)$, given by Eq. (5.22), over all space when occupied by only two molecules a distance R_{pq} apart, and subtracting the limiting form of that same integral as $R_{pq} \to \infty$. This potential is then used to calculate the binary cluster integral for a pair of molecules, denoted here as b' to distinguish it from the binary cluster integral for a pair of beads. The result is

$$b' = 2(\tfrac{1}{2} - \chi)\, \frac{\bar{v}^2 M^2}{V_S N_a{}^2}\, \mathscr{F}(X') \tag{5.25}$$

where

$$X' = 4C_M(\tfrac{1}{2} - \chi)M^{1/2}/\alpha^3 \tag{5.26}$$

and

$$C_M = (27/2^{3/2}\pi^{3/2})\bar{v}^2/V_1 N_a(M/\langle L^2\rangle_0)^{2/3} \tag{5.27}$$

The function $\mathscr{F}(X')$ decreases monotonically from unity to about 0.4 as X' increases from zero to ten—the range of most physical interest. For small values of X', the following series for $\mathscr{F}$ is useful:

$$\mathscr{F} = 1 - X'/2!2^{3/2} + X'^2/3!3^{3/2} \tag{5.28}$$

The parameter C_M is a formidable looking quantity, but it was introduced because it is very nearly a constant for any polymer–solvent system (it is, however, a very weak function of temperature). The value of C_M for polystyrene in a solvent of a molar volume of about 100 cm³/mol is about 0.04. In a good solvent, say, $\chi = 0.0$, $\alpha^3 X'$ increases from 12.6 to 126 as M increases from 10^5 to 10^7. For a poor solvent, say, for $\chi = 0.49$, $\alpha^3 X'$ increases from 0.25 to 2.51 over the same range of molecular weights.

The binary cluster integral given in Eq. (5.25) gives the following expression for π/C which is good only to the first power in C:

$$(\pi/C) = (RT/M)[1 + (M\bar{v}^2/V_1)(\tfrac{1}{2} - \chi)\mathscr{F}(X')C + \cdots] \tag{5.29}$$

The factor $F(X') \to 1$ as $X' \to 0$. It turns out, therefore, that only small corrections to Eq. (5.21) are required if X' is sufficiently small. For that reason, $\mathscr{F}(X')$ is sometimes referred to as the dilute solution correction to the Flory–Huggins theory. This interpretation is incorrect. The McMillan–Mayer approach is entirely different than that taken by Flory and Huggins in their original attack on polymer thermodynamics. Indeed the McMillan–Mayer approach gives only the first few virial coefficients, whereas the Flory–Huggins theory gives the free energy of mixing in closed form.

It is through the second virial coefficient that the parameters of the perturbation methods described in Section 5.1 can be related to those of the Flory–Huggins model. Zimm and Stockmayer[31] determined the second virial coefficient of polymer solutions using the perturbation technique and the potential of mean force described in Section 5.1. If the two approaches are to be compatible, the relationship between the binary cluster integral and the parameters used in the Flory–Huggins theory must be

$$b = 2(\tfrac{1}{2} - \chi)(\bar{v}^2/V_1 N_a)M^2 \qquad (5.30)$$

which gives

$$Z = (4/3^{3/2})C_M(\tfrac{1}{2} - \chi)M^{1/2} \qquad (5.31)$$

5.3. More on the Perturbation Method

The above relationships were obtained by equating the second virial coefficients derived from two different theoretical models for polymer–polymer and polymer–solvent interactions. We will see, however, that Eqs. (5.30) and (5.31) come out of the perturbation theory directly if the expression for the energetic and combinatorial effects of mixing lengths of chain and solvent is given by Eq. (5.22). In order to compute the potential of mean force, it is necessary to assume a form for the mean density of FH segments within the beads. It will be shown, however, that the outcome of the analyses is virtually independent of the nature of that segment density.

We will begin by assuming that each bead is a spherical region of radius R_0 containing the length of polymer chain comprising one statistical segment. We will also assume that, when averaged over all beads, the segment density is a constant everywhere within the beads. It follows from this last assumption that

$$R_0^2 = (3/5)\langle S^2 \rangle_\blacksquare \qquad (5.32)$$

where

$$\langle S^2 \rangle_\blacksquare = \text{mean-square radius of gyration of the length of}$$
$$\text{chain defined as one statistical segment} \qquad (5.33)$$

Since there are n statistical segments per molecule,

$$\langle S^2 \rangle_0 = n \langle S^2 \rangle_\blacksquare \qquad (5.34)$$

We will denote the molecular weight of a statistical segment as $M_\blacksquare$, where

$$M_\blacksquare = M/n \qquad (5.35)$$

The form of F_{pq} is found by integrating Eq. 5.22 over the volume occupied by the two beads when they are a distance R_{pq} apart (remembering that v_2 at any point has contributions from both beads) and subtracting the value of the same integral when $R_{pq} > 2R_0$. The result is

$$(F_{pq}/kT) = 0 \qquad\qquad \text{if } R_{pq} > 2R_0$$
$$(F_{pq}/kT) = (3\pi^{1/2}/5^{3/2})X(2 - 3h + h^3) \qquad \text{if } R_{pq} \leq 2R_0 \tag{5.36}$$

where

$$X = (3^{3/2}/4\pi^{3/2})(\bar{v}^2 M_{\bullet}^2/V_1 N_a \langle S^2 \rangle_{\bullet}^{3/2})(\tfrac{1}{2} - \chi) \tag{5.37}$$

and

$$h = (R_{pq}/2R_0) \tag{5.38}$$

When the expression for F_{pq} given in Eq. (5.36) is substituted into Eq. (5.12) and the integration carried out, the results can be expressed as

$$b = (8\pi^{3/2}/5^{3/2})R_0^3 X(1 - 0.1540X + \cdots) \tag{5.39}$$

One can carry out the same exercise using different assumed forms for the density of FH segments within the beads. If, for example, it is assumed that the density of segments falls off from the centers of mass in a Gaussian manner Eq. (5.39) is replaced by

$$b = (8\pi^{3/2}/5^{3/2})R_0^3 X(1 - 0.1768X + \cdots) \tag{5.40}$$

The factor R_0 in the above expression is defined by Eq. (5.32) and is introduced in this form only so one can make a direct comparison between Eqs. (5.39) and (5.40).

It is important to note that X is analogous to the parameter X' defined by the Flory–Krigbaum theory. There is, however, no dependence on an expansion factor since there is no excluded volume effect for the chain lengths making up a statistical segment; only short-range effects determine the Gaussian parameter of statistical segments. The other difference is that X' is proportional to $\sqrt{M}$ rather than $\sqrt{M_{\bullet}}$. It follows, therefore, that

$$X = \alpha^3 X'/\sqrt{n} \tag{5.41}$$

As a consequence, $X \ll 1.0$ for most real polymer–solvent systems, but could be as large as 0.5 for very high molecular weights (say, 10^7) and good solvents (say, $\chi = 0.0$). For most polymer–solvent systems, however, the series in parentheses in Eqs. (5.39) and (5.40) can be set equal to unity. Given the expression for R_0 and X, Eq. (5.32) gives

$$b = 2(\tfrac{1}{2} - \chi)(\bar{v}^2/V_1 N_a)M_\bullet^2 \qquad (5.42)$$

and

$$Z = [2/(4\pi)^{3/2}](\bar{v}^2/V_1 N_a)(M/\langle S^2 \rangle_0)^{3/2}(\tfrac{1}{2} - \chi)M^{1/2} \qquad (5.43)$$

which are, of course, the same equations as (5.30) and (5.31).

5.4. Expansion Factor over the Complete Range of Z

Although the series expressions for α_S^2 and α_L^2 are important in explaining the existence of the theta state for flexible macromolecules in solution, they converge too slowly to be useful in predicting chain expansion as functions of Z (assuming that they are functions of Z alone over its entire range of physically realistic values). There has thus been a continuing search for the solution to the intramolecular excluded volume problem —both numerically and in (at least approximate) closed form.

Flory[32] was the first to introduce a useful expression for the expansion factor. At that time, there was no distinction between expansion in S and L so no subscript was appended to the alpha. His expansion factor is, however, best interpreted as characterizing expansion in end-to-end distance, so we will refer to it as α_L.

The derivation given here is a bit more general than that originally presented by Flory, but the result is the same. We begin by assuming that the distribution of FH segments about the center of mass of the macromolecules is Gaussian, but of the form

$$\varrho = r\left(\frac{(\sqrt{6}\beta/\sqrt{\pi})^3}{\alpha_x \alpha_y \alpha_z}\right) \exp\{-6\beta^2[(\hat{x}/\alpha_x)^2 + (\hat{y}/\alpha_y)^2 + (\hat{z}/\alpha_z)^2]\} \qquad (5.44)$$

where the hatted coordinates are distances relative to the center of mass. Within the framework of this model, the α_i are molecular expansion factors and serve as a measure of the expansion (or compression) of individual molecules of the assembly over the unperturbed mean-square dimensions. All of the molecules for which $\alpha_x = 1$ have a value of $\langle S_x^2 \rangle$ equal to the unperturbed value for the assembly, with the same interpretation applying for the α_y and α_z. All of the molecules for which all three of the molecular expansion factors are equal to unity have a value of $\langle S^2 \rangle$ equal to the unperturbed value for the assembly.

The mean free energy of mixing of all of those polymer molecules with given values of α_x, α_y, and α_z can then be determined by integrating $\delta(\Delta F/kT)$ as given in Eq. (5.22) over all space surrounding one molecule with a FH segment density given by Eq. (5.44). The result to the first power in v_2 is

$$\Delta F/kT = (-1 + \chi) + (B/\alpha_x\alpha_y\alpha_z) \qquad (5.45)$$

where

$$B = 2C_M(\tfrac{1}{2} - \chi)\sqrt{M} \qquad (5.46)$$

If the probabilities of observing values of α_x, α_y, and α_z were given by the free energy of mixing alone the distribution functions characterizing these parameters would be proportional to $\exp[(1 - \chi)r - (B/\alpha_x\alpha_y\alpha_z)]$. But expansion of the molecular chains alters their conformational statistics and thus the entropy of the system. Flory suggested an appropriate measure of this entropy effect was that given by rubber elasticity. In other words, he assumed that the entropy effect of expansion of the chains in solution was the same as that for extending their end-to-end distance by three orthogonal extension factors equal to the α_x, α_y, and α_z. For a solution with $\tilde{N}$ polymer molecules per unit volume, incorporating the entropy effect of rubber elasticity gives the following for the distribution function of the expansion factors:

$$W = \text{const}(\alpha_x\alpha_y\alpha_z)^{\tilde{N}} \exp[-(\tilde{N}/2)(\alpha_x{}^2 + \alpha_y{}^2 + \alpha_z{}^2) - (\tilde{N}B/\alpha_x\alpha_y\alpha_z)] \qquad (5.47)$$

In principle, the expansion factors are obtained from the distribution W by performing the following integration:

$$\langle \alpha_x{}^2 \rangle = \frac{\displaystyle\int_{-\infty}^{+\infty} \alpha_x{}^2 W(\alpha_x{}^2, \alpha_y{}^2, \alpha_z{}^2)\, d\alpha_x\, d\alpha_y\, d\alpha_z}{\displaystyle\int_{-\infty}^{+\infty} W(\alpha_x{}^2, \alpha_y{}^2, \alpha_z{}^2)\, d\alpha_x\, d\alpha_y\, d\alpha_z} \qquad (5.48)$$

In this notation, the $\langle \alpha_L{}^2 \rangle$ is equivalent to $\alpha_L{}^2$ as defined in the perturbation theory. In order to obtain an approximate closed-form solution for the expansion factor, Flory suggested that the expectation values for the α_i be approximated by the values of α_i that maximize the probability distribution W. This is equivalent to assuming that W is sharp enough to be approximated by a delta function. Although this could hardly be justified mathematically, it does indeed give the expansion factor in simple closed form. Symmetry conditions require that the value of each of the three expansion factors that maximize W be equal. If we denote that value simply as α_L

(which is going back to the previous notation), it is easy to show that this value is given by solving the following simple equation

$$\alpha_L{}^5 - \alpha_L{}^3 = B = 2C_M(\tfrac{1}{2} - \chi)\sqrt{M} \tag{5.49}$$

Since the relationship between B and Z has already been established, Eq. (5.49) can be rewritten as

$$\alpha_L{}^5 - \alpha_L{}^3 = 2.60Z \tag{5.50}$$

This original Flory approach was, of course, extraordinarily oversimplified. Not only was the model itself an oversimplification, the mathematical treatment of it was approximate. To be rigorous, one should integrate over the distribution of α_i's to determine the expectation value rather that take that value of α_i that maximizes the distribution. Nonetheless, Flory captured the essential aspects of the phenomenon and expressed the results in the simple form of equation 5.49.

Stockmayer[33] pointed out that if the perturbation and Flory models of the excluded volume problem were to be consistent as $Z \to 0$, the coefficient in Eq. (5.50) was too large by almost exactly a factor of 2. He thus suggested using the following modified equation:

$$\alpha_L{}^5 - \alpha_L{}^3 = 1.33Z \tag{5.51}$$

Equation (5.51) is, of course semiempirical. There has thus been a continuing study of the excluded volume problem in an attempt to refine our understanding of the effect of polymer–solvent interaction.

One can improve on the Flory approach by performing the integration described in Eq. (5.48) instead of finding the value of the expansion factor that maximizes the distribution. Before we go into detail, however, we will make the problem more specific. Since it is the radius of gyration that is related to experimental observations, we will focus on the effect of the intramolecular excluded volume effect on its distribution and means. In the same fashion as Flory, we will assume that the assembly of macromolecules can be replaced by an assembly of segment clouds with densities given by Eq. (5.44). This is equivalent to assuming that fixing the radius of gyration of a chain completely determines its free energy of mixing. This is clearly an approximation; one could imagine many chain shapes all consistent with a given radius of gyration but with different free energies of mixing. How good or bad this preaveraging assumption is can only be determined by comparing its predictions with experimental results and more rigorous theoretical analyses.

We should note that further refinement can be introduced by using a more correct expression for the distribution of segments about the center of mass, but that subject will not be taken up here.

The definition of the parameters α_i and the assumed distribution of segments about the center of mass allow S_x, S_y, S_z and the polar radius S to be written explicitly in term of α_x, α_y, α_z, and α, where

$$\alpha^2 = (\alpha_x{}^2 + \alpha_y{}^2 + \alpha_z{}^2)/3 \qquad (5.52)$$

The first calculations of expansion factors using the preaveraged free energies of mixing were published in 1963 and were in reasonable agreement with known behavior of α at that time.[34] The distribution of radii of gyration are now, however, known very accurately. We will look at reevaluation of the expansion factor in light of our current knowledge of $W_{S_i}(S_i)$ and $W_S(S)$.

We will begin by writing the expression for $\langle \alpha_x{}^2 \rangle$ as

$$\langle \alpha_x{}^2 \rangle = \frac{\displaystyle\int_{-\infty}^{+\infty} \alpha_x{}^2 W(\alpha_x) W(\alpha_y) W(\alpha_z) \exp(-\Delta FM/kT)\, d\alpha_x\, d\alpha_y\, d\alpha_z}{\displaystyle\int_{-\infty}^{+\infty} W(\alpha_x) W(\alpha_y) W(\alpha_z) \exp(-\Delta FM/kT)\, d\alpha_x\, d\alpha_y\, d\alpha_z} \qquad (5.53)$$

Equation (5.53) gives $\Delta F_M/kT$ as a function of the three α_i. If this expression is expanded as a Taylor series in $(\alpha_x \alpha_y \alpha_z)^{-1}$ the result is

$$\exp(\Delta F_M/kT) = 1 - B/(\alpha_x \alpha_y \alpha_z) + \cdots \qquad (5.54)$$

If we ignore problems of convergence, both the numerator and denominator, Eq. (5.53) can be evaluated by integrating the series term-by-term. For our purposes, however, we need only evaluate the terms to the first negative power of the α_i. Since evaluation of the numerator and denominator depends on the existance of all negative moments of $W(\alpha_i)$, questions of convergence cannot be ignored. Although it is well known that all of the distribution functions in question approach zero extraordinarily rapidly as $S_i \rightarrow 0$ (they go to zero as $S_i{}^n$), this is not sufficiently fast to ensure the existance of negative moments greater than the nth. We must remember, however, that mathematical functions we use for $W(S_i)$ are all approximations of nature. Real macromolecules have distributions of radii of gyration that are always equal to zero below some critical value of S; real molecules cannot be compressed to a point! It must follow, therefore, that, *physically*, all of the negative moments must exist. It also follows that $\langle \alpha_x^2 \rangle$ can always be written as a Taylor series in B—although we will not concern ourselves with how rapidly such a series might converge.

Integrating the numerator and denominator in Eq. (5.53) and expressing the result as a Taylor series in B gives

$$\langle \alpha^2 \rangle = 1 + 0.5482B + \cdots \qquad (5.55)$$

or in terms of Z

$$\langle \alpha^2 \rangle = 1 + 1.424Z + \cdots \qquad (5.56)$$

Considering the rather uncertain effects of preaveraging the free energy of mixing, Eq. (5.56) is gratifyingly close to the accepted expression

$$\langle \alpha^2 \rangle = 1 + 1.276Z + \cdots \qquad (5.57)$$

Indeed, preaveraging introduces an error of only 11% in the linear term in the series expansion of $\langle \alpha^2 \rangle$.

Equation (5.53) can also be used to predict the asymptotic behavior of $\langle \alpha^2 \rangle$ at large Z. The methods will not be described in here; but we note that a saddle-point calculation was published in 1963.[33] The result is

$$\langle \alpha_s^2 \rangle = (12/5\pi^2) + (6/\pi)^{2/5}B^{2/5}$$
$$= 0.243 + 1.201Z^{2/5} \qquad (5.58)$$

Yamakawa[34] reports an asymptotic form of

$$\langle \alpha_s^2 \rangle = 1.228Z^{2/5} \qquad (5.59)$$

Forsman and Hughes[33] suggested the following empirical expression that patches together the limiting expressions given in equations (5.9) and (5.58):

$$\alpha^5 - \alpha^3 = 1.58(1 - 0.39/\alpha^2)Z \qquad (5.60)$$

Equation (5.53) was integrated using the eigenvalues for a chain of 500 statistical segments.[35] The results are shown in Figure 9 along with the asymptotic forms given in Eqs. (5.59) and (5.60). In addition, the modified Flory equation is shown in the same figure. All of the results are in good agreement except Eq. (5.60). We have no suggestion as to why the discrepancy exists. It is clear, however that the Flory expression, although approximate in its origins, is quite adequate as an analytical expression for the results of the preaveraged potential theory. The same conclusions must be drawn from examining the same sorts of comparisons given by Yamakawa.[34]

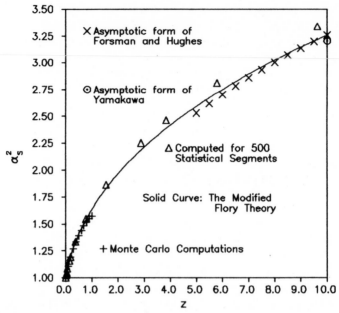

FIGURE 9. Expansion factor α_s^2 as a function of polymer–solvent interaction parameter Z for various theories.

6. THE DYNAMICS OF FLEXIBLE CHAIN MOLECULES

6.1. Linear Viscoelasticity

Polymer solutions and melts, i.e., polymeric liquids, demonstrate striking viscoelastic behavior—that is, they show mechanical properties intermediate between those of perfectly elastic solids and viscous fluids. These two limiting, idealized types of materials behave entirely differently in their response to deformation. An ideal elastic material stores all the energy of deformation whereas all energy expended in deformation of a fluid is dissipated as heat. These modes of behavior of idealized elastic solids and viscous liquids are demonstrated at all rates of loading. In contrast, viscoelastic materials store some of the energy of deformation and dissipate the rest, and the response is a function of the rate of loading. We often associate creep and stress relaxation with viscoelastic behavior. In addition, non-Newtonian viscosity of polymeric liquids is a manifestation of their viscoelastic nature.

The unique viscoelastic behavior of polymer solutions and melts is a consequence of the unique feature of their molecular architecture—their long, chainlike structure. Indeed it is the dynamics of the chainlike macro-molecules that controls the viscoelastic behavior of polymeric liquids.

For sufficiently small amplitudes of stress and strain, polymer liquids are linear viscoelastic. That is, measures of the amplitudes of the stress and strain histories are proportional, and are related through time functionality. For our purposes we will characterize linear viscoelastic behavior in terms of two measurable quantities, stress relaxation modulus and dynamic mechanical modulus. We will define the former in terms of the following hypothetical experiment: The polymer solution or melt will be subjected to an instantaneous shear strain Γ. While that deformation is maintained, the shear stress T decreases monotonically with time. The decrease in stress is described by the time-dependent stress relaxation modulus $G(t)$, the relationship being given by the expression

$$T(t) = G(t)\Gamma \tag{6.1}$$

The limit of $G(t)$ as $t \to \infty$ may be zero or some asymptotic nonzero value. The dynamic mechanical modulus is defined as follows: We will imagine that the polymer liquid or melt is subjected to a steady-state sinusoidal shear deformation described as

$$\Gamma^* = \Gamma_0 e^{i\omega t} \tag{6.2}$$

where Γ_0 is the amplitude of the deformation and ω is the angular frequency. The asterisk is used to indicate that the strain is a steady-state sinusoidal quantity. The response of the material is a steady-state sinusoidal stress

$$T^* = T_0 e^{i(\omega t + \delta)} \tag{6.3}$$

where δ is the phase angle between the stress and strain. The dynamic modulus G^* is then defined by the equation

$$T^* = G^*\Gamma^* \tag{6.4}$$

It can be written as the sum of real and imaginary parts as

$$G^* = G' + iG'' = G_0 e^{i\delta} \tag{6.5}$$

where

$$G_0 = [G'^2 + G''^2]^{1/2} \tag{6.6}$$

and

$$\tan \delta = G''/G' \tag{6.7}$$

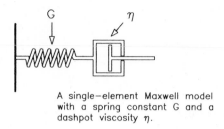

A single—element Maxwell model
with a spring constant G and a
dashpot viscosity η.

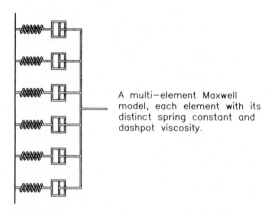

A multi—element Maxwell
model, each element with its
distinct spring constant and
dashpot viscosity.

FIGURE 10. Single- and multiple-element Maxwell models.

Dynamic mechanical measurements are particularly useful for characterizing the viscoelastic behavior of polymeric liquids. Since $G'' = 0$ for an ideal elastic solid, and $G' = 0$ for a Newtonian fluid, the phase angle can be used to define whether a material is more solidlike or more fluidlike. If δ is near zero, the material demonstrates mechanical properties very like an elastic solid; if δ is near 90°, it behaves very nearly as a viscous liquid.

A variety of mathematical models are useful in describing viscoelastic behavior. The Maxwell model is commonly used for describing the behavior of viscoelastic materials. As it turns out, the theory of chain dynamics presented in this chapter gives the behavior of polymeric liquids directly in terms of that model. To apply the Maxwell model is to assume that the stress–strain behavior of a material follows the same mathematical relationship as the force-deformation behavior of a spring and dashpot in series as shown in Figure 10. If a material behaved as a single Maxwell element with a spring constant of G and dashpot viscosity η, its viscoelastic behavior would be as follows:

$$G(t) = Ge^{-t/\tau} \tag{6.8}$$

$$G'(\omega) = G\omega^2\tau^2/(1 + \omega^2\tau^2) \tag{6.9}$$

and

$$G''(\omega) = G\omega\tau/(1 + \omega^2\tau^2) \tag{6.10}$$

where

$$\tau = \eta/G \tag{6.11}$$

The viscoelastic behavior of few, if any, real materials can be adequately described by one Maxwell element. In principle, however, it would be possible to describe the viscoelastic behavior of any material to any degree of accuracy by taking a sufficient number of Maxwell elements (including the possibility that the number of elements may have to approach infinity). For a multiple element Maxwell model with moduli G_i, viscosities η_i and relaxation times τ_i (where $\eta_i = G_i\tau_i$),

$$G(t) = \sum G_i e^{-t/\tau_1} \tag{6.12}$$

$$G'(\omega) = \sum G_i\omega^2\tau_i^2/(1 + \omega^2\tau_i^2) \tag{6.13}$$

and

$$G''(\omega) = \sum G_i\omega\tau_i/(1 + \omega^2\tau_i^2) \tag{6.14}$$

where the summations are taken over all of the elements. Since the viscosity of the multiple element model is the sum of element viscosities, the viscosity, η, of the viscoelastic liquid being modeled is written

$$\eta = \sum G_i\tau_i \tag{6.15}$$

6.2. Chain Dynamics[17]

The linear viscoelastic behavior of polymer systems derives from the dynamics of polymer chains. The model for macromolecular chains is the same as that we examined in considerable detail in Section 5. Each chain will consist of $n + 1$ beads connected by n statistical segments. The chain molecules will be assumed to be unperturbed. If W_0 is the joint distribution of the x_i, y_i, and z_i coordinates of the n statistical segments, it will be the product of the distribution functions of each statistical segment as given in equation (4.1). It is written

$$W_0(\mathbf{r}_1, \mathbf{r}_2, \ldots, \mathbf{r}_n) = \left(\frac{\sqrt{n}\beta}{\sqrt{\pi}}\right)^{3n} \prod \exp[-n\beta^2(x_i^2 + y_i^2 + z_i^2)] \tag{6.16}$$

where the vectors $\mathbf{r}_i$ represent the three orthogonal coordinates of the respective statistical segments.

Deformation of the polymeric liquid—either a step deformation as in stress relaxation or continuous deformation as in sinusoidal straining or steady flow—will effect a displacement of the macromolecules from their equilibrium conformational distribution. The joint distribution function for the components of the statistical segments will thus take on a form different from that given in Eq. (6.16), and will be a function of time. That nonequilibrium distribution will be denoted as $W(\mathbf{r}_1, \mathbf{r}_2, \ldots, \mathbf{r}_n, t)$. For the sake of simplicity, however, we will denote the equilibrium and nonequilibrium distributions as W_0 and W. Whenever chains are not distributed according to the equilibrium distribution W_0, there will be an entropic driving force for the chains to return to their random flight statistics. This driving force can be interpreted as a springlike effect acting on the beads through the statistical segments. Statistical mechanics gives a value for the x component of the force developed in the ith statistical segment as $kT[\partial \ln(W/W_0)/\partial x_i]$. The entropy spring force acting on bead i through statistical segments $i - 1$ and i is thus

$$kT\frac{\partial \ln(W/W_0)}{\partial x_i} - kT\frac{\partial \ln(W/W_0)}{\partial x_{i-1}} \tag{6.17}$$

To extend the model to include chain dynamics, we must also introduce a measure of the drag experienced by the macromolecular chains as they move in response to the entropy string effects. This measure will be in the form of a hydrodynamic drag coefficient, f, that gives a force on each bead that is proportional to the difference between its velocity and the velocity of the medium in which it is immersed. For the purpose of analysis, we will assume that the molecules are imbedded in a continuum fluid in pure shear. The velocity of the continuum fluid at the center of mass of the macromolecule will be taken as v_{S0}. If the shear rate is $\dot{\gamma}$, the velocity of the fluid at the position of the ith bead is $v_{S0} + \dot{\gamma}y$. For the moment we will assume that the motion of beads does not perturb the motion of the shear field of the continuum fluid. If f is the friction factor, the hydrodynamic force on a bead is

$$f(v_{S0} + \dot{\gamma}\hat{y}_i - \dot{\hat{x}}_i) \tag{6.18}$$

where $\dot{x}_i$ is the x component of the velocity of the ith bead. The force balance on each bead is thus written

Bead	Force balance

1    $f(v_{S0} + \dot{\gamma}\hat{y}_1 - \hat{\dot{x}}_1) + kT\dfrac{\partial \ln(W_0/W)}{\partial x_1}$ $= 0$

2    $f(v_{S0} + \dot{\gamma}\hat{y}_2 - \hat{\dot{x}}_2) + kT\dfrac{\partial \ln(W_0/W)}{\partial x_2} - kT\dfrac{\partial \ln(W_0/W)}{\partial x_1} = 0$

3    $f(v_{S0} + \dot{\gamma}\hat{y}_3 - \hat{\dot{x}}_3) + kT\dfrac{\partial \ln(W_0/W)}{\partial x_3} - kT\dfrac{\partial \ln(W_0/W)}{\partial x_2} = 0$

4    $f(v_{S0} + \dot{\gamma}\hat{y}_4 - \hat{\dot{x}}_4) + kT\dfrac{\partial \ln(W_0/W)}{\partial x_4} - kT\dfrac{\partial \ln(W_0/W)}{\partial x_3} = 0$

$\vdots$

n $f(v_{S0} + \dot{\gamma}\hat{y}_n - \hat{\dot{x}}_n) + kT\dfrac{2\ln(W_0/W)}{2x_n} - kT\dfrac{\partial \ln(W_0/W)}{\partial x_{n-1}} = 0$

$n+1$ $f(v_{S0} + \dot{\gamma}\hat{y}_{n+1} - \hat{\dot{x}}_{n+1})$ $\qquad - kT\dfrac{\partial \ln(W_0/W)}{\partial x_n} = 0$

$$(6.19)$$

If we subtract equation j from equation $j + 1$, the center-of-mass coordinates will be eliminated in terms of the x components of the statistical segments. The results are

Statistical segment	Force balance

1    $f(\dot{\gamma}y_1 - \dot{x}_1) + kT\dfrac{\partial \ln(W_0/W)}{\partial x_2} - 2kT\dfrac{\partial \ln(W_0/W)}{\partial x_1}$ $= 0$

2    $f(\dot{\gamma}y_2 - \dot{x}_2) + kT\dfrac{\partial \ln(W_0'/W)}{\partial x_3} - 2kT\dfrac{\partial \ln(W_0/W)}{\partial x_2}$
$\qquad\qquad + kT\dfrac{\partial \ln(W_0/W)}{\partial x_1} = 0$

3    $f(\dot{\gamma}y_3 - \dot{x}_3) + kT\dfrac{\partial \ln(W_0/W)}{\partial x_4} - 2kT\dfrac{\partial \ln(W_0/W)}{\partial x_3}$
$\qquad\qquad + kT\dfrac{\partial \ln(W_0/W)}{\partial x_2} = 0$

4    $f(\dot{\gamma}y_4 - \dot{x}_4) + kT\dfrac{\partial \ln(W_0/W)}{\partial x_5} - 2kT\dfrac{\partial \ln(W_0/W)}{\partial x_4}$
$\qquad\qquad + kT\dfrac{\partial \ln(W_0/W)}{\partial x_3} = 0$

$\vdots$

n $f(\dot{\gamma}y_n - \dot{x}_n)$ $\qquad\qquad - 2kT\dfrac{\partial \ln(W_0/W)}{\partial x_n}$
$\qquad\qquad + kT\dfrac{\partial \ln(W_0/W)}{\partial x_{n-1}} = 0$

$$(6.20)$$

Equations (6.20) can be written in matrix form as follows:

$$f(\dot{\gamma}\mathbf{y} - \dot{\mathbf{x}}) - kT\mathbf{A}\left[\frac{\partial \ln(W/W_0)}{\partial x_i}\right] = 0 \qquad (6.21)$$

where

$$\left[\frac{\partial \ln(W/W_0)}{\partial x_i}\right] = \begin{bmatrix} \dfrac{\partial \ln(W/W_0)}{\partial x_1} \\[2mm] \dfrac{\partial \ln(W/W_0)}{\partial x_2} \\[2mm] \vdots \\[2mm] \dfrac{\partial \ln(W/W_0)}{\partial x_n} \end{bmatrix} \qquad (6.22)$$

$$\mathbf{A} = \begin{bmatrix}
2 & -1 & 0 & 0 & 0 & \cdots & & \cdot & & \cdot \\
-1 & 2 & -1 & 0 & 0 & \cdots & & \cdot & & \cdot \\
0 & -1 & 2 & -1 & 0 & \cdots & & \cdot & & \cdot \\
0 & 0 & -1 & 2 & -1 & \cdots & & \cdot & & \cdot \\
\vdots & \vdots & \vdots & \vdots & \vdots & \vdots & \vdots & \vdots & \vdots & \vdots \\
0 & 0 & 0 & \cdot & \cdot & \cdots & -1 & 2 & -1 \\
0 & 0 & 0 & \cdot & \cdot & \cdots & 0 & -1 & 2
\end{bmatrix} \qquad (6.23)$$

and $\mathbf{y}$ and $\dot{\mathbf{x}}$ are column vectors of y_i and $\dot{x}_i$, respectively.

As they stand, Eqs. (6.21) cannot be solved because the ith equation contains derivatives of $\ln(W/W_0)$ with respect to x_i, x_{i-1}, and x_{i+1}. We note, however, that $\mathbf{A}$ is symmetric so we can effect an orthogonal transformation to a set of new coordinates—normal coordinates—in which the equations are separated. We thus introduce the transformation matrix $\mathbf{B}$ and normal coordinates $\boldsymbol{\xi}$ and $\boldsymbol{\zeta}$ where

$$\mathbf{x} = \mathbf{B}\boldsymbol{\xi}, \qquad \mathbf{y} = \mathbf{B}\boldsymbol{\zeta} \qquad (6.24)$$

The transformation relationship for derivatives gives

$$\left[\frac{\partial \ln(W/W_0)}{\partial x_i}\right] = \mathbf{B}\left[\frac{\partial \ln(W/W_0)}{\partial \xi_i}\right] \qquad (6.25)$$

The set of equations in (6.21) then becomes

$$f(\dot{\gamma}\mathbf{B}\boldsymbol{\zeta} - \mathbf{B}\boldsymbol{\xi}) - kT\mathbf{A}\mathbf{B}\left[\frac{\partial \ln(W/W_0)}{\partial \xi_i}\right] = 0 \qquad (6.26)$$

or

$$f(\dot{\gamma}\boldsymbol{\zeta} - \boldsymbol{\xi}) - kT\mathbf{B}^{-1}\mathbf{A}\mathbf{B}\left[\frac{\partial \ln(W/W_0)}{\partial \xi_i}\right] = 0 \qquad (6.27)$$

Since $\mathbf{A}$ is symmetric,

$$\mathbf{B}^{-1}\mathbf{A}\mathbf{B} = \boldsymbol{\Lambda} \qquad (6.28)$$

where $\boldsymbol{\Lambda}$ is diagonal. The matrix equation separates into the set of equations

$$f(\dot{\gamma}\zeta_i - \xi_i) - kT\lambda_i\left[\frac{\partial \ln(W/W_0)}{\partial \xi_i}\right] = 0 \qquad (6.29)$$

For linear chains the eigenvalues λ_p are given by

$$\lambda_p = 4\sin^2[p\pi/2(n+1)] \qquad (6.30)$$

It is beyond the scope of this chapter to present the details of the solution to the above set of equations. The physical interpretation of the outcome of the analysis is, however, easy to present. The instant the polymer solution or melt is subjected to the shear deformation, the chain molecules store energy by rubber elasticity. The energy is evenly divided between the normal modes, with each contributing exactly kT per molecule per unit volume to the shear modulus. If there are $\tilde{N}$ molecules per unit volume of solution or melt, the instantaneous stored energy is $\tilde{N}kT$ per unit of volume of solution. In terms of macroscopic units, the instantaneous shear modulus of the solution or melt is CRT/M.

At the instant the chain molecules are deformed, Brownian motion begins to drive the distribution function W back toward random-flight statistics. As chain motion takes place, each normal mode loses its stored energy at an exponential rate and the shear stress decreases. The shear modulus thus relaxes as the sum of exponentially decaying terms, i.e., as a Maxwell model with n elements, one for each statistical segment. The stress relaxation modulus is thus given by

$$G(t) = \frac{CRT}{M}\sum e^{-t/\tau_p} \qquad (6.31)$$

where the index p runs from 1 to n. The relaxation times are given by

$$\tau_p = f/\lambda_p 2kTn\beta^2 \qquad (6.32)$$

and the viscosity by

$$\eta = \frac{CRT}{M} \sum \tau_p \qquad (6.33)$$

We thus see that, just as in the case of the radius of gyration, the viscosity is the sum of contributions from each normal mode. Indeed, since the relaxation times decrease with increasing order (i.e., value of p), the most important contributions are from the lower order normal coordinates.

6.3. Interpretation of the Rouse Theory in Molecular Terms

In the continuum limit (that is, as $n \to \infty$), the orthogonal transformation and normal coordinates given in Eqs. (6.24) and (6.28) correspond to the normal coordinate analysis of the vibrating string. In that classical problem we learn that any deflection of the string from its equilibrium position can be described mathematically as the sum of changes in values of members of a set of normal coordinates. The same applies to changes in shape of a chain molecule. Any change in shape can be thought of as being due to the addition of contributions from changes in the normal coordinates. For example, consider the two-dimensional chain shown in Figure 11. All of the x_i and y_i coordinates will have certain values we will denote as $x_i^{(1)}$ and $y_i^{(1)}$. The orthogonal transformations given in Eqs. (6.24) then fix the normal coordinates at values $\xi_i^{(1)}$ and $\zeta_i^{(1)}$. We will now let the molecule change shape by changing the normal coordinates to a new set of values $\xi_i^{(2)}$ and $\zeta_i^{(2)}$. For the sake of illustration, we will let

$$\xi_i^{(2)} = \xi_i^{(1)} + c \qquad \text{and} \qquad \zeta_i^{(2)} = \zeta_i^{(1)} + c \qquad (6.34)$$

where c is some arbitrary quantity.

We will examine how a change in the value of each normal coordinate affects the shape of the chain by changing one at a time while leaving the rest at their initial values. Figure 11 shows the effect of setting $\xi_1^{(2)}$ equal to $\xi_1^{(1)} + c$ and $\zeta_1^{(2)}$ equal to $\zeta_1^{(1)} + c$. The result looks very much like a rigid rotation about the centermost element of the chain (it is not quite). The important point, however, is that a change in the first normal coordinate effects movement of the chain in blocks that are half length of the chain molecules while the centermost unit remains fixed.

If we do the same exercise with the second normal mode, Figure 11 shows that the effect is a motion of the chain in blocks that are 1/3 the length of the entire chain while the segments dividing the chain into the blocks remain fixed. That is, the fixed points serve as nodes for the defor-

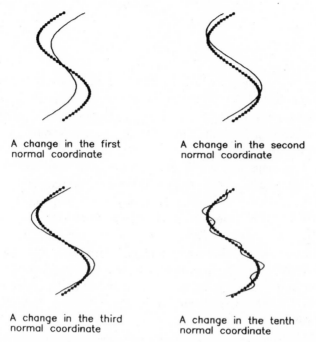

A change in the first
normal coordinate

A change in the second
normal coordinate

A change in the third
normal coordinate

A change in the tenth
normal coordinate

FIGURE 11. The effect of changing the first, second, third, and tenth normal coordinates on the shape of a chainlike molecule. The initial shape is represented by the chain of filled circles.

mation. Likewise, exciting the third normal mode effects motion about three nodes of blocks of chain that are 1/4 of the length of the entire chain. The significance of the normal modes in the Rouse analysis is straight-forward—the relaxation time associated with normal coordinate p corresponds to the reorganization of the chain conformation in blocks (or modes) of length $n/(p + 1)$. One of the more subtle aspects of the results is that the nodes are at the center of a bead or at a position midpoint between two beads, depending upon the order of the normal mode and whether n is even or odd.

It is also useful to examine the relative magnitude of the relaxation times themselves. If p is much less than n, the sine term in Eq. (6.30) can be set equal to its argument and we can write the following approximate equations:

$$\lambda_p = \frac{p^2\pi^2}{(n + 1)^2} \tag{6.35}$$

$$\tau_p = \frac{fn}{4\pi^2 k T \beta^2 p^2} \tag{6.36}$$

and

$$\tau_p = \frac{\tau_1}{p^2} \qquad (\text{if } p \ll n) \tag{6.37}$$

The approximate expression for τ_p will hold to within 4% if $p \leq (n/5)$ and to within 10% if $p \leq (n/3)$; the exact value of τ_n is 2.5 times greater than the value given by Eq. (6.37). Figures 12–14 show the predictions of the Rouse theory (in arbitrary units) for chains of 400 statistical segments comparing the use of the approximate and exact relaxation times. We note that there is a range of about a decade in ω over which G' and G'' are equal. For values of ω in this range and lower the approximate relaxation times are quite adequate. For values of ω greater than this, a significant error is introduced by using the equations (6.35)–(6.37).

The frequency range over which G' and G'' are approximately equal is called the transition zone, because it is the transition between the terminal zone, where viscous phenomena dominate, and the approach to the glassy zone, in which the Rouse model is totally inapplicable. The slope of log–log plots in the transition zone is almost exactly $\frac{1}{2}$. Indeed, it can be shown

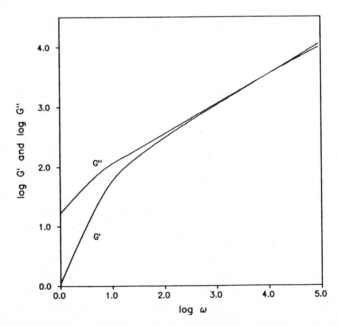

FIGURE 12. The real and imaginary parts of the dynamic modulus as a function of angular frequency ω computed from the exact eigenvalues of a linear, 400 statistical segment chain. Units are arbitrary.

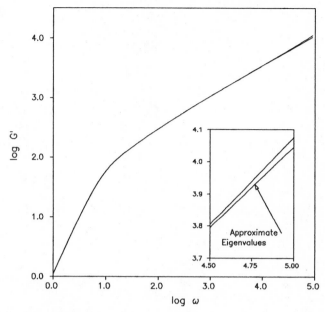

FIGURE 13. The real part of the dynamic modulus as a function of angular frequency ω computed from the exact and approximate eigenvalues of a linear, 400 statistical segment chain (the only discernible difference on the scale of the figure is at the high-frequency end of the transition zone). Units are arbitrary.

that in this viscoelastic domain the following equations adequately describe G' and G'' as functions of ω:

$$G'(\omega) \approx G'(\omega) \approx 0.55\tilde{N}kT(\tau_1\omega)^{1/2}$$
$$\approx 0.55\tilde{N}kT(fn\omega/kT4\pi^2\beta^2)^{1/2} \qquad (6.38)$$

It is useful to recast Eq. (6.38) into a form that best reflects experimental parameters. In particular, n and the friction factor f are inappropriate parameters, since they depend on the number of repeat units that make up a statistical segment. We note, however, that the product nf is the drag coefficient of the total molecule. If, then, we define a new quantity called the monomeric friction coefficient, ν, by the relationship

$$\nu M/M_0 = fn \qquad (6.39)$$

ν is independent of how the chain is divided up into statistical segments. Using the relationships $\tilde{N} = CN_a/M$, and $\langle S^2 \rangle = KM$, Eq. (6.38) can be

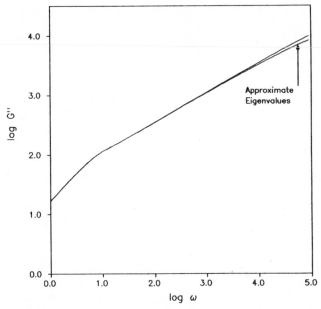

FIGURE 14. The imaginary part of the dynamic modulus as a function of angular frequency ω computed from the exact and approximate eigenvalues of a linear, 400 statistical segment chain (the only discernible difference on the scale of the figure is at the high-frequency end of the transition zone). Units are arbitrary.

transformed into

$$G' \approx G'' \approx \frac{2.2CN_a \nu KM}{M_0} \qquad (6.40)$$

Since the only parameter in Eq. (6.40) not measurable by independent experimental techniques is the monomeric friction factor, ν, it can be determined directly from viscoelastic behavior in the transition zone.

It is also useful to look again at the viscosity in light of what we know about the values of the lower-order eigenvalues and relaxation times as given in Eqs. (6.35)–(6.37). Substituting these values for τ_p into Eq. (6.33) gives

$$\eta = \tilde{N}kT \sum \frac{fn^2}{4\pi^2 kT\beta^2 p^2} \qquad (6.41)$$

$$= \frac{\tilde{N}fn}{4\pi^2\beta^2} \sum \frac{1}{p^2} \qquad (6.42)$$

We note that the summation in Eq. (6.42) is the same as that in Eq. (4.44) —and these summations appear as they do for the same reason; they represent the sum of contributions from the normal coordinates. If we make the approximation that p runs to ∞, the value of the summation is $\pi^2/6$. If we substitute this value into Eq. (6.42) and use the fact that $\beta^2 = 1/4\langle S^2 \rangle$, the viscosity for this limiting case is thus

$$\eta = \frac{\tilde{N}f\langle S^2 \rangle}{6} \tag{6.43}$$

According to the Rouse model, the viscosity of a solution or melt should be proportional to the number of chains per unit volume, the local frictional coefficient and the mean-square radius of gyration of the chains:

$$\eta = \frac{C\nu KN_a M}{6} \tag{6.44}$$

As simple as Eq. (6.44) is, it does work under rather restricted conditions that will be described in Chapter 3.

6.4. Limitations of the Rouse Theory

The Rouse theory is based on the random-flight model. A restriction in that model is that we be able to take enough repeat units per statistical segment to guarantee that they be stochastically independent and Gaussian, but that there be few enough that inertial effects—and now hydrodynamic drag effects—can be assumed to act through point masses (i.e., the beads). Aside from these two restrictions, the number of repeat units per statistical segment is arbitrary. As a consequence, any theoretical predictions that are expressed as functions of n must be suspect. If predictions do turn out to be functions of the number of statistical segments, we are forced to incorporate n as an empirical, adjustable parameter. Equation (6.36) shows that τ_p is a function of n only if p is sufficiently large. Or, in other words, the longest relaxation times are independent of n, but for normal modes of order greater than some critical value (which would be set by practical considerations) the dependence on n creeps in with increasing p. For example, if we take a chain of 100 statistical segments, Eq. (6.36) is accurate (good to within 4%) up to $p = 20$. This implies that the relaxation times are independent of n up to τ_{20}, which is equal to $\tau_1/400$. This implies that Rouse-like behavior can be depended upon for about two and a half decades in relaxation time. For a chain of 1000 statistical segments, Rouse-like behavior is predicted for over four and a half decades of relaxation time. Clearly the range in relaxation time over which one can expect the Rouse predictions to apply increases markedly with chain length—or with molecular weight for real macromolecules.

Two of the assumptions in its development severely restrict the applicability of the Rouse theory. It was assumed that the chains were in their unperturbed state and that the motion of the beads did not perturb the flow field of the continuum in which the polymer molecules were embedded. The solution must be at least semidilute for these two assumptions to apply. First, it is now well accepted that the excluded volume effect vanishes as the polymer concentration becomes great enough for the distribution of repeat units to be a constant throughout the solution. Furthermore, if that criterion is met, every element of volume that is large compared to $\langle S^2 \rangle_{\blacksquare}^{3/2}$ but small compared to $\langle S_2 \rangle_0^{3/2}$ must be in the same hydrodynamic state as every other, and each bead will see the same shear field. Although this satisfies the assumption that the velocity of the continuum is equal to $v_{S0} + \dot{\gamma} y_i$, the continuum must now be considered a mixture of solvent and repeat units and not just solvent as first assumed by Rouse.

It is this last restriction on the Rouse theory that introduces the most troublesome aspect of interpreting experimental data on the linear viscoelasticity of polymer solutions in terms of theory. If this theory is restricted to solutions that are at least semidilute, it is restricted to solutions in which there is extensive interpenetration of the chain molecules. At the same time, however, it was assumed that each chain acts independently—i.e., that no intermolecular cooperative effects are required for the polymeric fluid to respond to stress. Such cooperative effects are treated in Chapter 3.

We should conclude this section by pointing out that Zimm[35] formulated the same problem solved by Rouse[17] in the position coordinates of the beads themselves. The result is the same, but the Zimm approach is required if hydrodynamic bead–bead interactions are to be incorporated into the theory. This is necessary for treating chain dynamics in dilute solution. Since, however, we will be focusing on concentrations in the semidilute regime and greater, we will not pursue this topic further.

7. GRAPH THEORY AND THE STATISTICS AND DYNAMICS OF RANDOM-FLIGHT CHAINS

Up to this point, this chapter has treated chain statistics and chain dynamics as two different subjects, although it was pointed out that the same matrix was involved in the solution of conformational statistics and dynamics problems for linear chains. In this section, however, we will show that the two theoretical developments are intimately related through some of the simplest notions of graph theory.

7.1. The Graph Theoretical Description of Branched Random-Flight Chains

Only the most elementary notions from graph theory are required in the description of the model.[36] For the purposes of this chapter, we need only consider a graph as a collection of points connected by lines. In general, the lines can run between the points in any way whatsoever. Indeed, any number of lines—or no lines at all—can radiate from any of the points. If the lines are assigned directions, i.e., can be represented as vectors, the graph is called a *directed* graph. No closed loops (rings) will be considered in this development, so all the graphs will have n vectors and $n + 1$ points. Such graphs are referred to as "trees." Figure 15 shows a directed graph with eight points (shown as circles for clarity) and seven vectors. The points (circles) and vectors have been numbered in an arbitrary way.

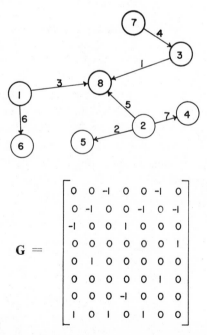

$$
G = \begin{bmatrix}
0 & 0 & -1 & 0 & 0 & -1 & 0 \\
0 & -1 & 0 & 0 & -1 & 0 & -1 \\
-1 & 0 & 0 & 1 & 0 & 0 & 0 \\
0 & 0 & 0 & 0 & 0 & 0 & 1 \\
0 & 1 & 0 & 0 & 0 & 0 & 0 \\
0 & 0 & 0 & 0 & 0 & 1 & 0 \\
0 & 0 & 0 & -1 & 0 & 0 & 0 \\
1 & 0 & 1 & 0 & 1 & 0 & 0
\end{bmatrix}
$$

FIGURE 15. A directed graph of eight points (represented by circles) and seven vectors, and its incidence matrix **G**.

We see that there is a one-to-one correspondence between a directed graph and our pictorial representation of random-flight chains—except that we have, so far, considered only linear chains and numbered the masses and statistical segments in a sequential fashion. We will now, however, consider random-flight chains, with any type of branching and with their beads and vectors numbered in any arbitrary way, as being represented by such a directed graph.

Graph theory teaches that for every directed graph of the type shown in Figure 15 there is an associated matrix $\mathbf{G}$, the incidence matrix of the vectors, in which

$$
\begin{aligned}
G_{ij} &= -1 \quad && \text{if vector } j \text{ starts at mass } i \\
&= +1 \quad && \text{if vector } j \text{ terminates at mass } i \\
&= 0 \quad && \text{otherwise} \quad\quad\quad\quad\quad\quad (7.1)
\end{aligned}
$$

Figure 15 shows the associated matrix along with the graph representation of the chain.

7.2. Chain Statistics

Equations (4.5) give the three orthogonal components of the polar radius of gyration as functions of the x, y, and z components of the position vectors of the beads in the bead-spring model. In that section of this chapter we were considering linear chains, and it was assumed the beads were numbered successively from one end to the other. It is clear, however, that Eqs. (4.5) hold as well for any kind of branched molecules, and regardless of how the masses are numbered. Since there are only $3n$ degrees of freedom in the definition of chain conformation—regardless of whether we are dealing with branched or linear chains—one of the $\hat{x}_i$, $\hat{y}_i$, and $\hat{z}_i$ must be eliminated by the relationships

$$
x_k = - \sum_{\substack{i=1 \\ \neq k}}^{n+1} \hat{x}_i
$$

$$
y_k = - \sum_{\substack{i=1 \\ \neq k}}^{n+1} \hat{y}_i \quad\quad\quad\quad (7.2)
$$

$$
z_k = - \sum_{\substack{i=1 \\ \neq k}}^{n+1} \hat{z}_i
$$

The predictions of any theory of chain behavior must, of course, be independent of which set of coordinates is eliminated, so k is arbitrary. We learned in Section 4 that the sums of squares of the types given in Eqs. (4.5) can be written as

$$\sum_{i=1}^{n+1} \hat{x}_i^2 = \hat{x}^T \mathbf{M} \hat{x} \tag{7.3}$$

with equivalent expressions for the $\hat{y}$ and $\hat{z}$ components, where $\hat{x}$ is the column vector of the remaining n coordinates and $\mathbf{M}$ is an $n \times n$ matrix that is independent of which coordinate was eliminated, and where M_{ij} is equal to 2 if $i = j$ and equal to 1 otherwise.

In order to analyze chain statistics, it is convenient to transform from $\hat{x}_i$ coordinates to the x_i, the components of the statistical segments. One can write

$$x_i = \hat{x}_j - \hat{x}_k \tag{7.4}$$

where subscripts j and k are determined by i, the choice of numbering systems and how the vectors are directed in the graph representation. For the example given in Figure 15

$$x_2 = \hat{x}_5 - \hat{x}_2 \tag{7.5}$$

Equation (7.4) thus represents a linear transformation

$$x = \mathbf{P}\hat{x} \tag{7.6}$$

where $\mathbf{P}$ is an $n \times n$ matrix with elements determined by the graph representation of the chain. Matrix $\mathbf{P}$ for linear chains is given in Eq. (4.18). Given the above expression relating x and $\hat{x}$, Eq. (7.3) then reduces to

$$S_x^2 = (n+1)^{-1} x^T (\mathbf{P}^{-1})^T \mathbf{M} \mathbf{P}^{-1} x \tag{7.7}$$

$$= (n+1)^{-2} x^T \mathbf{F} x \tag{7.8}$$

where, just as for linear chains,

$$\mathbf{F} = (n+1)(\mathbf{P}^{-1})^T \mathbf{M} \mathbf{P}^{-1} \tag{7.9}$$

The connection between the $\mathbf{P}$ and $\mathbf{G}$ matrices is easily established. Examination of the definition of $\mathbf{G}$ shows that

$$x = \mathbf{G}^T \hat{x}' \tag{7.10}$$

where $\hat{x}'$ is a column matrix containing all $n + 1$ coordinates of the masses.

Note that, since $\mathbf{G}^T$ is an n by $(n + 1)$ matrix, the column vector x has n elements as required, but has contributions from all $n + 1$ values of $\hat{x}_i$. Equation (7.10), as it stands, is therefore inappropriate for application to either chain statistics or dynamics. We must replace $\hat{x}'$ with the vector $\hat{x}$ in which one value of $\hat{x}_i$, say, $\hat{x}_r$, has been replaced by the expression

$$\hat{x}_r = - \sum_{\substack{i=1 \\ i \neq r}}^{n+1} \hat{x}_i \tag{7.11}$$

This can be accomplished by the transformation

$$\hat{x}' = \mathbf{T}\hat{x} \tag{7.12}$$

where $\mathbf{T}$ is an $(n + 1)$ by n matrix, $\hat{x}$ is an n-order column vector with $\hat{x}_r$ missing, and

$$\begin{aligned} T_{ij} &= \delta_{ij}, && \text{if } j < r \\ &= -1, && \text{if } j = r \\ &= \delta_{i-1,j}, && \text{if } j > r \end{aligned} \tag{7.13}$$

The net result is

$$x = \mathbf{G}^T \mathbf{T} \hat{x} \tag{7.14}$$

or

$$\mathbf{P} = \mathbf{G}^T \mathbf{T} \tag{7.15}$$

and

$$\mathbf{F} = (n + 1)(\mathbf{T}^T\mathbf{G})^{-1}\mathbf{M}(\mathbf{G}^T\mathbf{T})^{-1} \tag{7.16}$$

Eq. (7.16) is an awkward expression relating $\mathbf{F}$ and $\mathbf{G}$. In a subsequent section we will establish the far simpler expression

$$\mathbf{F} = (n + 1)(\mathbf{G}^T\mathbf{G})^{-1} \tag{7.17}$$

Inverting Eq. (7.14) and writing

$$\hat{x} = \mathbf{Q}x \tag{7.18}$$

where

$$\mathbf{Q} = \mathbf{P}^{-1} \tag{7.19}$$

shows that the n coordinates $\hat{x}_i$ that are taken as the independent position variables are linear functions of the n statistical segments x_i. Since each x_i is distributed in a Gaussian fashion with a distribution parameter $\sqrt{n}\beta$, each independent $\hat{x}_i$ variable will be distributed in a Gaussian fashion with a parameter given by

$$\beta_i^2 = n\beta^2 \left[\sum_{j=1}^{n} Q_{i,j}^2 \right]^{-1} \qquad (7.20)$$

An interesting peculiarity of the above development is that Eq. (7.20) does not apply to the position coordinate $\hat{x}_r$ that was eliminated in the definition of the n independent position variables. This is no problem, of course, since the position coordinate that is eliminated is quite arbitrary, and clearly the distribution of *every* bead about the center of mass is Gaussian. The best way out of the dilemma is the following: It follows from Eqs. (7.12) and (7.18) that we can write

$$\hat{x}' = \mathbf{H}x \qquad \text{where } \mathbf{H} = \mathbf{TQ} = \mathbf{T(G}^T\mathbf{T})^{-1} \qquad (7.21)$$

so that the expression

$$\beta_i^2 = n\beta^2 \left[\sum_{j=1}^{n+1} H_{i,j}^2 \right]^{-1} \qquad (7.22)$$

gives the Gaussian parameters for all of the $(n + 1)\hat{x}_i$ coordinates.

The total mean segment density about the center of mass, ϱ, is thus given by

$$\varrho = \sum_{j=1}^{n+1} \varrho_i \qquad (7.23)$$

where ϱ_i is the Gaussian distribution of $\hat{x}_i$, and has the same form as Eq. (4.1), with β_i replacing $\sqrt{n}\beta$. It must be stressed, however, that the distributions ϱ_i and ϱ are distributions about the center of mass of the chains, and represent the probability of finding a given bead (or, in the case of ϱ, any bead) in an interval dx—or, in the three dimensional extension of the above equations, in a volume element $dV = dx\,dy\,dz$. Indeed, the 3-dimensional version of Eq. 7.23 is the correct distribution which is approximated by Eq. 5.44.

7.3. The Distance Matrix and Scattering Phenomena

The scattering of light, x-rays, and neutrons has played and important role in polymer science. Theoretical details of these scattering phenomena are beyond the scope of this chapter. Chapter 4 does, however, go into considerable detail on polymer characterization by neutron scattering. Nonetheless, some elementary (and somewhat oversimplified) scattering relationships will be introduced so they may be related to the material on chain statistics already discussed.

Consider the scattering of light or neutrons from point scattering centers located in a volume V, which will be referred to as the scattering volume. The intensity of scattered radiation relative to the intensity of the incident beam as a function of angle is characterized by a scattering function $S(Q)$. The experimental parameter Q, is given by

$$Q = (4\pi/\lambda) \sin(\theta/2) \tag{7.24}$$

where λ is the wavelength of the radiation and θ is the angle between the incident and scattered beams. Although Q itself is often referred to as the scattering vector, it is, in fact, the magnitude of a vector $\mathbf{Q}$ which is defined such that its direction is the same as that of the difference between unit vectors directed along the incident and scattered beams.

Information about the spatial relationship between the various scattering centers within the scattering volume is given by the expression

$$S(Q) = \int e^{-i\mathbf{R}\cdot\mathbf{Q}} \langle \varrho_s(\mathbf{R})\varrho_s(\mathbf{R} + \mathbf{R}') \rangle \, d\mathbf{R} \tag{7.25}$$

which is the three-dimensional Fourier transform of the density autocorrelation function of the scattering centers. In this notation, $\mathbf{R}$ is a position vector and $d\mathbf{R}$ is the differential in volume. As for all Fourier transforms, the integral is taken from $-\infty$ to $+\infty$ in each of the three orthogonal directions. But since the density autocorrelation function is identically zero outside of V, one can think of the integration as being taken only over the scattering volume. The autocorrelation function is given by

$$\langle \varrho_s(\mathbf{R})\varrho_s(\mathbf{R} + \mathbf{R}') \rangle = \int \varrho_s(\mathbf{R})\varrho_s(\mathbf{R} + \mathbf{R}') \, d\mathbf{R}' \tag{7.26}$$

where $\varrho_s(\mathbf{R})$ is the density of point scattering centers within V, and the integral is taken over the scattering volume.

We will assume that the scattering centers are the beads of an assembly of random flight chains, the centers of mass of which are uncorrelated in the scattering volume. For this special case the scattering function is designated as $P(Q)$ and is written

$$P(Q) = [2/n(n-1)] \int_{-\infty}^{\infty} W_{ij}(r_{ij})e^{-\mathbf{Q}\cdot\mathbf{r}_{ij}} \, dr_{ij} \tag{7.27}$$

where $\mathbf{r}_{ij}$ is the vector connecting beads i and j, W_{ij} is the Gaussian distribution for the end-to-end distance of that vector, and where the summation is taken over all of the $n(n - 1)/2$ distinct i–j pairs. Integrating Eq. (7.27) gives

$$P(Q) = [2/n(n - 1)] \sum e^{-(Q^2/4\beta_{ij}^2)} \tag{7.28}$$

The above expression for linear chains in the limit as $n \to \infty$ was evaluated in closed form long ago[37] and is written as

$$P(Q) = (2/q^4)(e^{-q^2} - 1 + q^2) \tag{7.29}$$

where q is a dimensionless "scattering vector" given by

$$q = Q\langle S^2 \rangle^{1/2} \tag{7.30}$$

Expressions for $P(q)$ for various other special cases have also been published.[3]

It is significant that the matrix relationships resulting from the graph theoretical representation of random-flight chains gives the required Gaussian parameters in a straightforward way for chains of any type of branching. We can apply Eq. (7.21) directly to obtain an expression for $\hat{x}_{ij}$, the x component of vector $\mathbf{r}_{ij}$. The result is

$$\hat{x}_{ij} = \hat{x}_i - \hat{x}_j = \sum_{k=1}^{n} (H_{ik} - H_{jk})x_k \tag{7.31}$$

We thus define a new $(n + 1)$ by $(n + 1)$ matrix, $\mathbf{W}$, which will be called the *distance matrix*, with elements

$$W_{ij} = \sum_{k=1}^{n} (H_{ik} - H_{jk}) \tag{7.32}$$

The elements of $\mathbf{W}$ are integers equal to the number to statistical segments that span the distance between beads i and j. All of the diagonal elements of $\mathbf{W}$ are, of course, equal to zero. Given the definition of the distance matrix, the Gaussian parameters for the vectors $\mathbf{r}_{ij}$ and the three orthogonal components are given by

$$\beta_{ij}^2 = n\beta^2/W_{ij} \tag{7.33}$$

and the scattering function thus becomes

$$P(Q) = [2/n(n - 1)] \sum e^{-(Q^2 W_{ij}/4n\beta^2)} \tag{7.34}$$

7.4. Chain Dynamics

Section 6 treated the Rouse[17] model in some detail and described how it gives a satisfactory explanation of the linear viscoelastic behavior of polymeric fluids under rather strict conditions of concentration and molecular weight. The formulation was, however, for linear chains alone. We show here, however, that the analysis can be extended to chains of any type of branching by exploiting the properties of the incidence matrices described earlier.

In order to make the notation simpler to work with than that used in Section 6, we will introduce the following symbolic equivalence:

$$Ex_i \equiv kT\left[\frac{\partial \ln(W/W_0)}{\partial x_i}\right] \tag{7.35}$$

The utility of this notation is that the product of the symbolic spring constant, E, times x_i, including the vector definition of direction defined in the directed graph, behaves under the types of transformations used in Section 6.2 exactly as the derivative in Eq. (7.35). As an example, Figure 16 shows that statistical segment k exerts an equal and opposite (average) springlike force Ex_k on beads i and j.

The following set of equations gives the familiar Rouse[17] analysis in its simplest form, i.e., one space dimension and zero continuum shear field:

$$\begin{aligned}
-f\dot{x}_1 + Ex_1 &= 0 \\
-f\dot{x}_2 + E(x_2 - x_1) &= 0 \\
-f\dot{x}_3 + E(x_3 - x_2) &= 0 \\
-f\dot{x}_4 + E(x_4 - x_3) &= 0 \\
&\vdots
\end{aligned} \tag{7.36}$$

which leads to the matrix equation

$$fx + \mathbf{A}Ex = \mathbf{0} \tag{7.37}$$

where $\mathbf{0}$ is a column vector with all elements equal to zero. The matrix $\mathbf{A}$ is matrix (6.23) which was discussed in Section 6.

Now we consider entropy forces in an arbitrarily branched assembly of chains. Statistical segment k will be assumed to connect beads i and j as shown in Figure 16. Any number of other statistical segments may originate or terminate at beads i and j, and vectors a, b, and c and l, m,

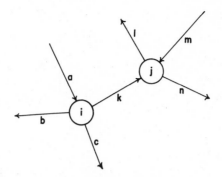

FIGURE 16. Statistical segment k acting as an entropy spring on beads i and j.

and n only serve as examples. The forces on the two beads can be written

$$F_i = E(x_k - x_a + x_b + x_c + \cdots) \tag{7.38}$$

$$F_j = E(-x_k + x_l - x_m + x_n + \cdots) \tag{7.39}$$

The number of terms appearing in each expression and their signs are thus determined by the number of vectors associated with each mass and their orientations.

Taking the difference between Eqs. (7.38) and (7.39) in the head–tail sense gives the entropic force acting on statistical segment k. This is the force term F_k applicable to the Rouse model and its extensions for branched chains. The result is

$$F_k = -E(2x_k - x_a + x_b + x_c - x_l + x_m - x_n \cdots) \tag{7.40}$$

The dynamic equation for an assembly of freely draining chains can therefore still be written as Eq. (7.37) [or (6.21)], but using the following set of rules for construction of matrix $\mathbf{A}$:

1. $A_{ii} = 2$
2. $A_{ij} = -1$ if vectors i and j meet in a head-to-tail configuration.
3. $A_{ij} = +1$ if vectors i and j meet in a tail-to-tail or head-to-head configuration.
4. $A_{ij} = 0$ if vectors i and j do not meet.

If we now consider $\mathbf{G}^T\mathbf{G}$ we find that it is an n by n matrix with elements defined exactly according to the above set of rules. It therefore follows that the Rouse matrix for both linear and branched random-flight chain is given by

$$\mathbf{A} = \mathbf{G}^T\mathbf{G} \tag{7.41}$$

a fact that has been misinterpreted in the past.[39]

Using the same type of arguments, it is equally simple to show that the analogous $(n + 1)$ by $(n + 1)$ matrix used in the Zimm[35] treatment of chain dynamics is given by

$$\mathbf{Z} = \mathbf{G}\mathbf{G}^T \tag{7.42}$$

where the elements of $\mathbf{Z}$ for linear chain are the same as for a corresponding $\mathbf{A}$, except for elements 1,1 and $(n + 1),(n + 1)$ which are replaced by unity.

7.5. Relationship between Chain Dynamics and Chain Statistics

In order to establish the relationship between chain dynamics and chain statistics we must prove Eq. (7.17) which, in light of the development given in the last section, can now be written

$$\mathbf{F} = (n + 1)\mathbf{A}^{-1} \tag{7.43}$$

In view of Eq. (7.16), this is equivalent to a proof that

$$\mathbf{G}^T\mathbf{T}\mathbf{M}^{-1}\mathbf{T}^T\mathbf{G} = \mathbf{G}^T\mathbf{G} = \mathbf{A} \tag{7.44}$$

In order to begin the proof, we must first consider the matrix $\mathbf{M}^{-1}$. Simple substitution shows that it is given by

$$\mathbf{M}^{-1} = \boldsymbol{\delta} - (n + 1)^{-1}\mathbf{1} \tag{7.45}$$

where $\mathbf{1}$ is an n by n matrix with all its elements equal to unity. If we define a matrix $\mathbf{R}$ by the equation

$$\mathbf{R} = \mathbf{T}\mathbf{M}^{-1}\mathbf{T}^T \tag{7.46}$$

then

$$R_{ij} = \sum_{k,l=1}^{n+1} T_{ik}[\delta_{k1} - (n+1)^{-1}]T_{j1} \tag{7.47}$$

$$= \sum_{l=1}^{n+1} T_{i1}T_{j1} - (n+1)^{-1} \sum_{k,l=1}^{n+1} T_{ik}T_{jk} \tag{7.48}$$

If r is the index of the bead coordinate eliminated by operating on x'

with **T**, the elements of the first summation of Eq. (7.48) are

$$
\begin{array}{ll}
0 & \text{if } i \neq j \neq r \neq i \\
+1 & \text{if } i = j \neq r \\
-1 & \text{if } i = r \text{ and } j \neq r \ \text{ or if } i \neq r \text{ and } j = r \\
n & \text{if } i = j = r
\end{array}
\tag{7.49}
$$

It is convenient to write this contribution of **R** as $\boldsymbol{\delta} + \mathbf{R}'$ where

$$
\begin{aligned}
(n+1)R'_{ij} &= 0 & \text{if } i \neq r \neq j \\
&= -(n+1) & \text{if } i = r \neq j \text{ or } i \neq r = j \\
&= +(n+1)(n-1) & \text{if } i = j = r
\end{aligned}
\tag{7.50}
$$

The second summation in Eq. (7.48) can be examined by defining a new matrix $\mathbf{R}''$ by the equation

$$
\mathbf{R}'' = \mathbf{T1T}^T
\tag{7.51}
$$

Performing this matrix multiplication gives

$$
\begin{aligned}
R''_{ij} &= 1 & \text{if } i \neq r \neq j \\
&= -n & \text{if } i = r \neq j \text{ or } i \neq r = j \\
&= n^2 & \text{if } i = r = j
\end{aligned}
\tag{7.52}
$$

Consequently,

$$
R'_{ij} - (n+1)^{-1}R''_{ij} = -(n+1)
\tag{7.53}
$$

or

$$
\mathbf{R} = \boldsymbol{\delta} - (n+1)^{-1}\mathbf{1}
\tag{7.54}
$$

Since it can be easily demonstrated that

$$
\mathbf{G}^T\mathbf{1} = \mathbf{0} \quad \text{and} \quad \mathbf{1G} = \mathbf{0}
\tag{7.55}
$$

where **0** represents a matrix with all elements equal to zero, the net result is

$$
\mathbf{G}^T\mathbf{R}\mathbf{G} = \mathbf{G}^T\mathbf{G} = \mathbf{A}
\tag{7.56}
$$

It has therefore been established that

$$
\mathbf{F} = (n+1)\mathbf{A}^{-1}
\tag{7.57}
$$

The eigenvalues of $\mathbf{A}$, λ_p, when incorporated into the Rouse[17] theory and its extensions, determine the dynamic behavior of the system of random-flight chains. Likewise, the eigenvalues of $\mathbf{F}$, Φ_p, determine the statistical properties. Equation (7.57) requires, however, that

$$\lambda_p = (n + 1)\Phi_p^{-1} \tag{7.58}$$

It follows, therefore, that the eigenvalues of $\mathbf{G}^T\mathbf{G}$ predict all of the statistical and dynamic properties of an assembly of unperturbed, branched random-flight chains in the absence of hydrodynamic interactions.

7.6. An Example

As an example of the application of the ideas presented in this section, we will consider the comblike chain of 36 statistical segments and 37 beads shown in Figure 17. The $\mathbf{G}$, $\mathbf{A}$, and $\mathbf{W}$ matrices are shown in Figures 18–20. The eigenvalues of the $\mathbf{F}$ matrix for linear and branched chains are listed in Table 2. The one-dimensional distribution of beads about the center of mass as given by Eq. (7.23) for linear chains of 36 statistical segments and branched chains of the type illustrated in Figure 17 are shown in Figure 21. The scattering functions for these two types of chains are given in Figure 22, and the real and imaginary parts of their dynamic moduli in Figures 23 and 24.

We will not consider the distribution of radius of gyration of branched chains, since calculations of similar examples have been published.[39,40] It is, however, a trivial exercise to obtain $\langle S^2 \rangle$ from Eq. (4.33). For the 36-segment chains considered in this example, the ratio of the mean-square radius of gyration of an assembly of the branched chains to that of an assembly of linear chains is 0.625.

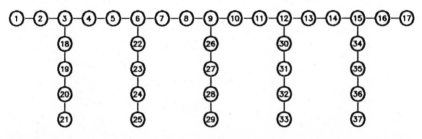

FIGURE 17. An example of a branched chain of 37 beads and 36 statistical segments. In the example discussed in the text, the corresponding directed graph was defined by stipulating that all of the vectors point either to the right or down.

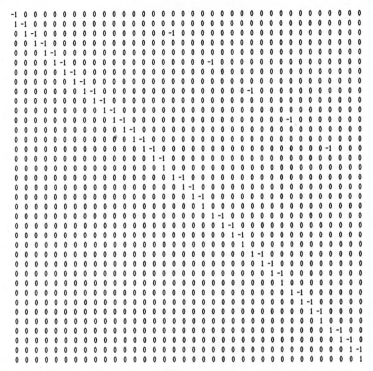

FIGURE 18. The **G** matrix for the branched 36-segment chain discussed in the text.

The effect of branching on the mean bead density is as one would expect—Figure 21 shows that the beads of the branched chains are more closely distributed about their center of mass.

It is clear from Figure 22 that a linear chain of 36 statistical segments does not demonstrate the same scattering function as an infinite chain—it falls off considerably faster with q for values of q greater than about 4. The chain length effect is interesting in itself, but it must be remembered that a 36-segment chain has too few statistical segments to describe the behavior of high-molecular-weight macromolecules. It would be instructive to examine the chain length effect for chains with $n > 200$. The important illustration in Figure 22 is, however, the marked effect of branching on scattering. One would need to do computations for chains with far larger numbers of statistical segments in order to generate scattering functions that could be used in the interpretation of experimental data. Nonetheless, the behavior shown in Figure 22 is a qualitative picture of how this type of branching might be expected to effect the scattering function of real macromolecules.

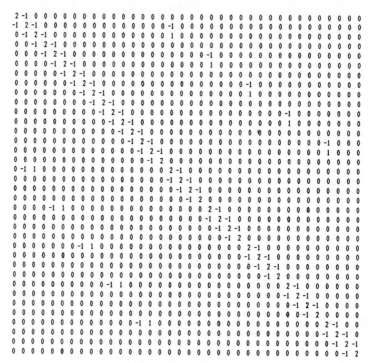

FIGURE 19. The **A** matrix for the branched 36-segment chain discussed in the text.

TABLE 2. Eigenvalues of the **F** Matrix for Linear (L) and Branched (B) Chains of 36 Statistical Segments[a]

L	B	L	B	L	B
5135.306	2374.229	33.649	37.000	12.137	13.300
1286.143	762.444	29.499	37.000	11.601	12.601
573.340	502.910	26.160	37.000	11.140	10.939
323.865	434.194	23.438	23.216	10.744	10.804
208.400	187.984	21.192	21.600	10.405	10.677
145.685	185.387	19.320	19.396	10.118	10.640
107.878	136.869	17.747	17.832	9.877	10.617
83.347	96.867	16.415	17.113	9.680	8.502
66.536	68.245	15.281	15.134	9.522	8.680
54.520	51.101	14.310	15.129	9.402	8.289
45.638	48.175	13.476	14.756	9.317	8.104
38.891	37.000	12.757	14.133	9.267	7.984

[a] Branched chain shown in Figure 17.

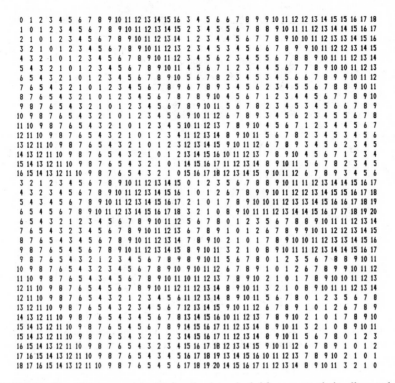

FIGURE 20. The distance matrix **W** for the branched 36-segment chain discussed in the text.

Since the illustration is for a chain of only 36 statistical segments, the computed G' and G'' do not go far into the transition zone with increasing ω before the relaxation times become physically meaningless. These viscoelastic functions are thus plotted in Figures 23 and 24 as functions of increasing ω only up to the low-frequency end of the transition zone. Nonetheless, the profound effect of branching on both energy storage and energy dissipation is clearly demonstrated.

Although we have demonstrated that various useful relationships come out of the graph theoretical representation of branched random-flight chains, we have not examined such complexities as excluded volume, coupling of the motion of solvent molecules with the motion of the beads, or cooperative effects in chain dynamics. Since, however, all of the spacial and statistical properties of the macromolecules can be written in terms of the matrix elements associated with their graph theoretical representation, one could predict that such extensions will be forthcoming.

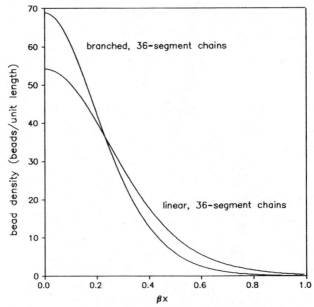

FIGURE 21. The one-dimensional density of beads about the center of mass for a linear chain of 36 statistical segments and the branched 36-segment chain discussed in the text.

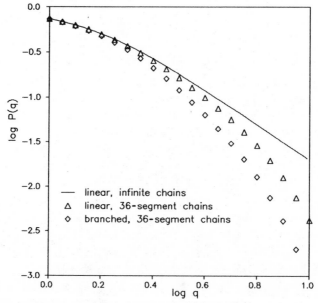

FIGURE 22. The scattering function $P(q)$ for an infinite chain, for a linear chain of 36 statistical segments, and for the branched 36-segment chain discussed in the text.

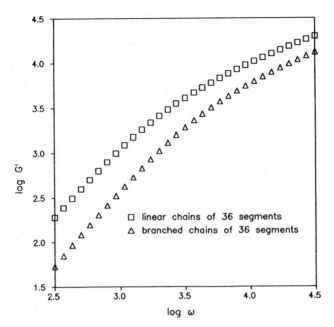

FIGURE 23. The real part of the dynamic modulus for a linear chain of 36 statistical segments and the branched 36-segment chain discussed in the text.

It should be noted that there have been many and varied approaches to the analysis of chain statistics, scattering phenomena, and chain dynamics. The unique feature of the method described in this chapter, and illustrated by the above example, is unity. One starting point, the incidence matrix of the graph theoretical representation of the random-flight chain, leads directly to all of the computed quantities.

In the not-too-distant past, limitations in computing speed and memory made applications of methods such as those described here far less attractive than today. The example given in this chapter was, however, worked out on a microcomputer in very reasonable times. Computing problems should no longer stand in the way of complete modeling of systems of flexible polymer molecules (that is, those that can be adequately described as random-flight chains). It seems reasonable to expect, therefore, that the graph theoretical considerations described in this chapter will be useful as tools for working toward that end.

No discussion of the application of graph theory to polymer systems would be complete without reference of Eichinger's elegant review of chain statistics.[38] The focus in that work is on applications of the Kirchhoff

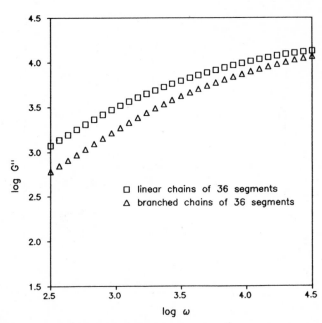

FIGURE 24. The imaginary part of the dynamic modulus for a linear chain of 36 statistical segments and the branched 36-segment chain discussed in the text.

matrix, which is a generalization to branched chains of the matrix introduced by Zimm in his analysis of chain dynamics.

REFERENCES

1. P. J. Flory, *Principles of Polymer Chemistry*, Cornell University Press, Ithaca, New York (1953).
2. H. Tompa, *Polymer Solutions*, Butterworths Scientific Publications, London (1956).
3. H. Yamqkawa, *Modern Theory of Polymer Solutions*, Harper and Row, New York (1971).
4. W. W. Graessley, *Adv. Polym. Sci.* **47**, 67 (1982).
5. P. G. de Gennes, *Scaling Concepts in Polymer Physics*, Cornell University Press, Ithaca, New York (1979).
6. J. W. Dettman, *Mathematical Methods in Physics and Engineering*, McGraw-Hill, New York (1962).
7. C. R. Wylie, *Advanced Engineering Mathematics*, McGraw-Hill, New York (1974).
8. A. Papoulis, *The Fourier Integral and Its Applications*, McGraw-Hill, New York (1962).
9. D. C. Champeney, *Fourier Transforms and their Physical Applications*, Academic Press, London (1973).

10. R. E. Walpole and R. H. Myers, *Probability and Statistics for Engineers and Scientists*, Macmillan, New York (1972).
11. H. Cramer, *The Elements of Probability Theory*, Wiley, New York (1955).
12. W. Kuhn, *Kolloid Z.* **76**, 258 (1936); *Kolloid Z.* **87**, 3 (1939).
13. M. V. Volkenstein, *Configurational Statistics of Polymeric Chains*, Interscience, New York (1963).
14. P. J. Flory, *Statistical Mechanics of Chain Molecules*, Interscience, New York (1969).
15. W. C. Forsman, *J. Chem. Phys.* **65**, 4111 (1976).
16. W. C. Forsman and R. A. Hughes, *J. Chem. Phys.* **38**, 2118 (1963).
17. P. E. Rouse, *J. Chem. Phys.* **21**, 1272 (1953).
18. M. Fixman, *J. Chem. Phys.* **36**, 306 (1962).
19. W. C. Forsman, *J. Chem. Phys.* **42**, 2829–2835 (1965).
20. R. F. Hoffman and W. C. Forsman, *J. Chem. Phys.* **50**, 2316–2324 (1969).
21. B. H. Zimm, W. H. Stockmayer, and M. Fixman, *J. Chem. Phys.* **21**, 1716 (1953).
22. M. Fixman, *J. Chem. Phys.* **23**, 1656 (1955).
23. M. Gordon, S. B. Ross-Murphy, and H. Suzuke, *Eur. Polym. J.* **12**, 733 (1975).
24. S. Aronowitz and B. E. Eichinger, *J. Polym. Sci., Polym. Phys. Ed.* **13**, 1655 (1975).
25. S. K. Gupta and W. C. Forsman, *Macromolecules*, **5**, 779-785 (1972).
26. J. E. Martin, *Macromolecules* **17**, 1263 (1984).
27. P. J. Flory, *Principles of Polymer Chemistry*, Chap. 12, Cornell University Press, Ithaca, New York (1953).
28. J. H. Hildebrand, *J. Chem. Phys.* **15**, 225 (1947).
29. P. J. Flory and W. R. Krigbaum, *J. Chem. Phys.* **18**, 1086 (1950); P. J. Flory, *J. Chem. Phys.* **17**, 1347 (1949).
30. W. G. McMillan and J. E. Mayer, *J. Chem. Phys.* **13**, 276 (1945).
31. W. H. Stockmayer, *J. Polym. Sci.* **15**, 595 (1955); W. H. Stockmayer, *Makromol. Chem.* **35**, 54 (1960).
32. P. J. Flory, *Principles of Polymer Chemistry*, Chap. 14, Cornell University Press, Ithaca, New York (1953).
33. W. C. Forsman and R. A. Hughes, *J. Chem. Phys.* **38**, 2123 (1963).
34. H. Yamqkawa, *Modern Theory of Polymer Solutions*, Chap. 3, Harper and Row, New York (1971).
35. B. H. Zimm, *J. Chem. Phys.* **24**, 256 (1956).
36. C. Berge, *The Theory of Graphs*, John Wiley and Sons, New York (1962).
37. P. Debye, *J. Phys. Colloid Chem.*, **51**, 18 (1947).
38. B. E. Eichinger, *Macromolecules* **13**, 1 (1980).
39. S. K. Gupta and W. C. Forsman, *Macromolecules* **6**, 285 (1973).
40. S. K. Gupta and W. C. Forsman, *Macromolecules* **7**, 853 (1974).

Dielectric and Electroviscous Properties in Flowing Polymer Systems

HERMANN BLOCK

1. INTRODUCTION

The dynamic behavior of polymer solutions and their relation to macro-molecular properties is most commonly investigated in terms of the study of viscosity or of dynamic birefringence. Of these techniques the latter provides the more direct information about macromolecular conformation and distortion in shear while the former, in terms of viscoelastic behavior in particular, gives perhaps more direct evidence on the normal mode relaxations in flow. As well as these dynamic techniques which apply to solutions being subjected to flow fields there are a wide range of techniques which are extensively applied in still solutions with the purpose of obtaining information on the conformation and molecular motion of macromolecules. However, these techniques have only infrequently been employed in situations where the solution is simultaneously subjected to a flow field. At least in principle, extra information characterizing macromolecules or even colloidal or particulate suspensions can result from investigating shear induced changes in some of these properties. Thus, changes in scattered light intensity[1] and in permittivity in flow have received some attention and it is the

HERMANN BLOCK • School of Industrial Science, Cranfield Institute of Technology, Cranfield, Bedford MK43 OAL, United Kingdom.

latter technique which is the subject of this chapter. Although no formal name has been coined for the study of permittivity of flowing liquids it is suggested that the term *flow modified permittivity* might be appropriate.

The study of complex permittivity ε^* in still solution provides, via its frequency dependence, information on both the dipole moment and relaxation rates for some of the normal modes of macromolecular coils in solutions. It is a well established investigative technique.[2] One can expect that since the net dipole moment of a polymer chain is the vectorial sum of the residue contributions,[2,3] changes in coil shape in shear would be reflected in alterations in the permittivity. Even in the absence of deformation under shear, changes in permittivity will occur because, for rotational flow, the space and time averaged distribution of dipoles is different from still solution, where only Brownian rotational diffusion leads to an averaging. In terms of frequency dependence, the study of flow-modified permittivity has close analogies to dynamic birefringence but with two very significant differences. Firstly, dynamic birefringence reflects flow anisotropy in polarization at optical frequencies and these frequencies are too high to reflect the rates of whole macromolecular rotation or even the internal modes involving conformational change. This need not be so for flow-modified permittivity. Secondly, flow-modified permittivity can be studied over a much wider range of frequencies (at least from 10^2 to 10^6 Hz). No such wide and conveniently selectable range of frequencies is practical in dynamic birefringence work.

In a practical sense also, flow-modified permittivity has a number of advantages over dynamic birefringence as well as some disadvantages. Much higher shear rates are possible since there is no requirement for light to traverse the shear gap and in consequence much narrower gaps are possible. Indeed, the sensitivity of the method is enhanced by working with rather narrow gaps since the electrical response, particularly in terms of capacitance, is increased. On the other hand, the technique has the disadvantage of all dielectric measurements in that polar solvents produce extra difficulty in measurement, although, as will be developed further in Section 5, conductance does not interfere to anywhere near the extent in dynamic permittivity as in static measurements.

Many systems exhibit changes in permittivity in flow and the phenomena are not limited to polymer solutions but include colloids and particulate suspensions.

It is convenient to classify polymer solutions in terms of polymer flexibility as rigid or flexible. The division of course is not absolute: at low enough flow rates quite flexible polymers do not distort in shear while polymers which are rigid at moderate shear rate may distort under extreme flow fields. The division stems from the consideration that a rigid polar macromolecule will produce changes in permittivity in flow which are entirely due to the shear field perturbing the spatial and time fluctuating distribution of Brownian motion. The average contribution of polarization per macromolecule is thus different from that predicted by, for example, the Debye–Pellat theory, and is given by theories such as those due to Saito and Kato[4] and Barisas.[5] Flexible polymers undergo more complex changes in flow since not only is the fluctuating spatial distribution influenced but other factors such as the summative dipole may change if the coil is distorted. Theory here is more complex and much less developed, and there is some evidence that other complicating factors such as changes in internal field with shear partake in the phenomenon. Such theoretical considerations are discussed in more detail below.

Rigid and flexible polymers show, at least for the systems studied up to date, markedly different effects. Particularly, rigid polymers tend to show a decrease in the real component of permittivity with shear while flexible coils show the reverse, although this may in part be due to the inclination of residue dipoles in the chain as discussed in Section 4.

The third group of macromolecular systems which show distinctive behavior are colloidal or particulate suspensions. These have their own internal polarization in an electric field and the shear field has influence only in spinning the entire polarizable entity. Here forced resonances are observed in the permittivity if the electric field frequency has values of the same order as the shear rate and this resonance phenomenon is discussed in Section 5. A result of the lag in polarization and the spinning of the particle is that the bulk viscosity of such a suspension is enhanced by an electric field. This phenomenon of an influence of electric field on viscosity was first discovered by Winslow[6,7] and although the effect has been attributed to a number of causes it is suggested that a common basis exists for the dielectric resonances and Winslow effect, a hypothesis discussed further in Section 7.

2. TECHNIQUES OF MEASUREMENT

Since at the present time no commercial flow apparatus is available for measuring flow-modified permittivity, equipment has to be constructed by the investigator. Certain general design features are important for studying changes in permittivity occurring in flowing liquids, the principle requirements being that both the electric and flow fields are well defined. This means that the electrodes should be smooth, parallel, and either planar or of low curvature so that a homogeneous field results, and that such electrodes should also serve as the shear-generating surfaces. Obviously capillary flow is not very suitable even if one electrode is a central wire and the other the walls of the capillary. Channel flow and flow between disks are possible arrangements, but the simplest and most effective in providing a large range of accessible shear rates is the coaxial cylinder system or Couette type of cell.

In the design of Couette cells there are a number of design features to which attention must be paid if artifacts of measurement are to be avoided. Some of these are common to the design considerations applicable to dynamic birefringence or viscosity measuring equipment based on Couette cells. Indeed there is no reason why constructed equipment should not be capable of measuring either or both of these properties as well as flow-modified permittivity. The design considerations which are important are as follows:

(i) The flow in the gap between the two cylinders should be laminar. Obviously surface roughness must be avoided, but even with smooth surfaces there is a limit in shear rate above which turbulence sets in. The condition for the onset of turbulence in the gap between coaxial cylinders, one of which is subject to rotation, has been extensively analyzed by Taylor.[8] At high enough shear rates laminar flow in such a situation cannot be maintained and turbulence, initially in the form of Taylor vortices, sets in. The onset point is determined by the radius of either cylinder and the gap between them. The larger the radius and the narrower the gap the higher the shear rate for the onset of turbulence. Also, especially at wider gaps, there is an advantage in rotating the outer rather than inner cylinder since the Coriolis forces increase the achievable rate of rotation before turbulence sets in. Any designer of a Couette cell should consult the data given by Taylor to determine the maximum shear rate for any proposed design. In practice it is probably difficult to achieve shear rates greater than $\sim 5 \times 10^5$ s^{-1} for shear-modified permittivity studies, and this limit is an order of magnitude larger than for dynamic birefringence work because of the gap required for the latter.

(ii) At high shear rates it is impossible to adequately thermostat the liquid in the gap when long times of shear are involved. The viscous resistance causes local heating, generating a thermal gradient in the gap which will affect all types of measurement. Champion[9] describes this problem in studying dynamic birefringence at high shear rates and overcomes it by severely restricting the time of rotation and thus of measurement. Although in the study of flow-modified permittivity a small temperature gradient does not have the catastrophic effect that ensues in dynamic birefringence (distortion or even loss of light due to thermal refraction), a similar limitation in the time available for measurement can occur. The presence of an unacceptable temperature increase can be monitored either with thermocouples imbedded in the electrodes near the surface or by monitoring the permittivity before and after a period of high shear rate.

(iii) Measurements of complex permittivity over the frequency range 10^2–10^6 Hz are made using radio frequency bridges in which the coaxial cylinder system acts as a more or less "lossy" capacitor. Changes in permittivity due to shear should not reflect distortions of any free liquid surface in flow. Rather than a sealed system we have overcome this problem by using three terminal measurements with a guard electrode sited through and below the free surface. Adequate screening of this nature is very important and it is probable that some early work mainly reflects changes due to meniscus distortion in flow.

(iv) Correct alignment on a common axis of both stator and rotor is an important requirement for accurate measurements, particularly of changes in permittivity. Not only does an out-of-center inner cylinder cause a nonuniform shear gradient and a reduction in the achievable shear rate before turbulence sets in, but any changes in its position results in spurious capacitance and conductance fluctuations. Machining and alignment of very close tolerance of the coaxial cylinders is required for equipment with gaps $\sim$0.02 mm. Two approaches have been taken in the cells used by us involving either freely suspended or semifixed stators. The former achieves self-centering under shear but has a lower limit of shear rate below which centering is lost. Semifixed stators overcome this problem at the expense of complexity in design and engineering requirements, particularly if concurrent viscosity measurements are to be made. It should be appreciated that the coaxial cylinders must be electrically insulated from each other with materials of low loss. Most insulators of this type which are readily machined have flexibility which can cause a flexing of the stator axis under shear, if it is not coincident with the center of the rotor. Viscosity measurements on self-centering or semifixed stators can be achieved by the use of a torque bar attached to the stator as described below.

A brief description of the important features of the self-centering and semifixed stator cells designed by the author are given below.

2.1. The Self-Centering Couette Cell[10]

Figure 1 shows some of the constructional details of the cell. The compound stator STC consisted of a brass electrode, its widest diameter electrically connected through a rod and suspension wire to an insulated terminal in the suspension-chuck PC. This was the live electrode. At the upper portion of the electrode, the suspension rod is covered with a PTFE sleeve and an outer brass sleeve carrying a fine coil of wire from its upper edge. This provides a guard electrode since the brass sleeve penetrates the surface of the liquid located just above the quartz annulus Q_2 which carries a short PTFE ring (to stop liquid flowing out into the cell housing). The stainless steel rotor R carries at its lower extremity a quartz disk Q_1 which is held in by an "O" ring and an extended locking cylinder EX. Quartz annulus Q_2 and disk Q_1 provide two transparent windows so that this cell can be used for dynamic birefringence measurements. Electrical contact to the rotor is made by means of a pin PN dipping into a mercury cup MC. There is no deliberate insulation of the rotor from its housing H but experience has shown that electrical contact via the bearings B, or the use of carbon brushes to contact the rotor leads to electrical noise, and that a mercury cup contact eliminates this. The mercury cup and housing are electrically shorted externally.

The rotor is driven by a variable speed dc motor (not shown) and belt BT. Inside the housing there is an oil-drainage system OD which enables both upper and lower bearings to be lightly and periodically oiled from the oil point OP. The whole housing is carried in a steel cradle C, slides on insulated guides TS and locates against the insulating block TF where it is held by Tufnol bolts BL. The housing was mounted on a vertical, wall-mounted optical bench and because of its metallic construction acted as an electrical screen. The rotation rate was measured by means of the slotted disk D (carrying three channels) and a lamp–photodiode system P which produced light pulses whose rate was counted by a pulse counter. An inlet pipe 'EP and A act as an entry system for thermostatted air, the end of A also locating and locking the oil drainage system OD.

Since for self-centering the suspension wire has to be thin and the torque forces on the stator are large, it is necessary to prevent the stator twisting the wire through large angles. To achieve this a silver steel pin Z is located in a slot SL1 machined in a polycarbonate block S. This pin in its slot prevents any large rotation of the stator. The lever LV moving in slot SL2 interacts with the pin Z and is connected to an electromagnetic balance. After calibration with standard liquids the torque developed can be

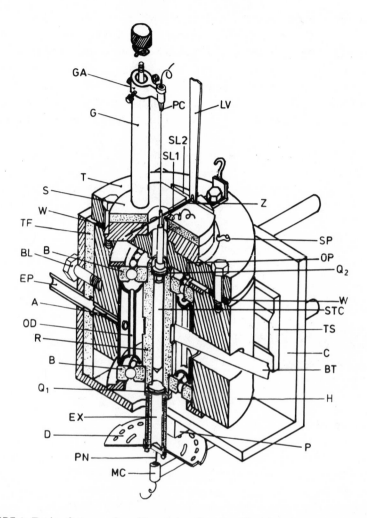

FIGURE 1. Design features of a self-centering Couette cell with facility for flow-modified permittivity, dynamic birefringence and viscosity measurements (taken from Block *et al.*[10]). R, rotor with belt BT; STC, suspended stator from pin chuck PC; Q_1, Q_2, quartz sealing disk and annulus; H, housing; B, bearings; BL, Tufnol bolt of which the other member is in the face removed for section; OD, oil drainage system with oil point OP and inlet pipe A acting as antirotation lock; T, table adjustable with lead washers W; S, stop system held by springs SP; G, gantry with screw adjustment GA and locking cap (shown expanded); C, cradle with Tufnol slides TS and flat TF; EP, air entry pipe; EX, lower compression ring carrying slotted disk D and contact pin RN; P, lamp–photodiode holder; PN, pin with MC, mercury cup; Z, silver steel pin in locating slot SL 1 and reacting lever LV in slot SL 2; HK, lifting hook.

used to measure the viscosity. The table T which can be leveled by means of compressing the lead washers W, carries this suspension system including the gantry G and screw adjustment GA for the suspension-chuck PC, and provides for this system some degree of centering. For narrower stators, used only in dynamic birefringence, this part of the assembly could be removed and a chuck with fixed stator placed on the table. The taper locates the gantry assembly or the alternative chuck, both of which are held in place by springs SP.

Table 1 gives some of the specifications of the self-centering Couette cell.

Although this apparatus is particularly effective at high shear rates ($\leq 3 \times 10^5$ s^{-1}) there is a lower limit which, for not too viscous liquids, is

TABLE 1. Specifications for Couette Cells[10,11]

a. Freely suspended stator cell

Rotor bore	1.586 cm
Rotor internal length	12.0 cm
Stator diameter	1.654 or 1.82 cm
Gap (capacitance estimated)	0.110 or 0.0228 mm at 20 °C
Effective stator length	6.35 cm
Effective stator area	31.2 or 36.3 cm²
Volume capacity of cell	~ 7 cm³

b. Semifixed stator cell

Rotor bore	4.445 cm
Stator diameter	4.430 cm
Maximum ovality in rotor and stator	2.54×10^{-4} cm
Gap	0.075 mm
Maximum run-out on precision bearings	2.54×10^{-4} cm
Length of S2 Electrode System	2.54 cm
Effective area of large electrode	35.3 cm²
Effective area of microelectrode	0.785 mm²
Diameters of rods P and Q	0.873 cm
Diameters of clearance holes for P and Q	0.900 cm
Volume capacity of cell	~ 10 cm³
Temperature range (upper limit not exceeded at present time)	(20–50) $\pm$ 0.1 °C

of the order of 10^3 s^{-1} and below which centering is difficult to achieve. In part the lower limit is determined by the care exercised in mounting the cell vertically. Certainly it is not possible to determine the zero shear rate capacitance and this has to be obtained by a process of comparison between shear and nonshear measurements (in a separate dielectric cell) using a simple liquid (such as toluene) which is not subject to flow modified permittivity effects.

Thermal stability both over long periods during thermostatting and short periods during shear was measured dielectrically as already described, but the system of air thermostatting using an external heat exchanger was not of high stability ($\pm 2\,^{\circ}$C) nor could temperatures $> 45^{\circ}$ be easily achieved. Torque measurements were not entirely satisfactory being on occasion subjected to noise of up to $\sim 20\%$.

2.2. Semifixed Stator Couette Cell[11]

In order to study shear-modified permittivity and the effect of electric fields on viscosity at lower shear rates a cell with at least a partially restrained coaxial cylinder system is required. We have recently constructed such a cell whose design is shown in Figure 2.

The major design problem with a fixed stator cell is to achieve a satisfactory method of measuring the viscosity at low shear rate since the torque developed on the inner cylinder is rather small. A fixed cylinder with strain gauge measurement of deformation is unlikely to be sensitive enough. Some flexibility of one cylinder in a twisting mode, but fixed alignment along the central axis, is the design principle chosen by us. The inner cylindrical stator consists of three parts S1, S2, and S3 with S2 aligned accurately with S1 and S3 by means of a pin bearing B. Two supporting rods at 180° (one shown in Figure 2 as P) fix the position of S3 to S1 and thus support S2 via the pin bearing, but the holes (shown as H) in S1 and S2 through which rods Q1, Q2, and P pass are oversize. The result is that S2 can rotate on its bearing to an extent limited by the gaps. At right angles to the support rods P the rods Q1 and Q2 (top of Q2 only shown) lead from S2 through the oversize holes in S1 to the upper surface of the cell and are locked onto the pressure block C. On rotation of the rotor R, the torque transmitted to the surface of S2 by liquid in the gap between rotor and stator causes a deflection of C and this is transmitted via pressure points D (one only shown) to a beam which for clarity is not shown in Figure 2, but which is rigidly fixed to the "U"-shaped block A at the screw points E. Either limb of block A can be used depending on the direction

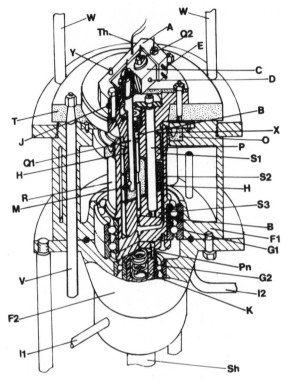

FIGURE 2. Design features of a semifixed stator cell with facility for flow-modified permittivity and viscosity measurements (taken from Block *et al.*[11]). S1, S2, and S3, semifixed stators of which S2 has limited rotation on bearings B; H, oversize holes in S1 and S2 permitting rotation; R, rotor with slot X; P, Q1, and Q2 rods of which Q1 leads to microelectrode M and Q2 to surface electrode of S2; C, pressure block with pressure point D; A, beam-carrying block with beam attachment screws E and lifting points J; G1 and G2, bearings; F1 and F2, housing; O, upper oil-seal; Sh, drive shaft via pin drive Pn; K, contact spring; I1 oil inlet and I2 oil outlet; T, stator table; V, guide rods; W, gantry rods for retracting mechanism (not shown); Th, thermocouple; Y, solution aspirating port.

of rotation; screws E serve as points of attachment. The flexing of the beam is measured by attached strain gauges and a selection of beam thicknesses enables a range of torques to be measured, and viscosities to be estimated.

Electrical measurements are made by having the inner surface of the rotor R as one electrode, the surface of S1 and S3 as guard electrode, and the surface of S2 as measuring electrodes. To this end the body of S2 is made of insulating glass-filled PTFE (all sections of insulators in Figure 2 are shown dotted) so that electrical contact between the measuring electrodes and the guard electrodes cannot occur via P, even on full rotation of S2. Rod Q2 serves as contact electrode to the surface metal annulus on S2 and Q1 goes to a small microelectrode M in the surface of S2; it is insulated from the main electrode by a PTFE annular insert. Thus two electrodes one of large and the other of small surface area (~ 1 mm²) can be used for measurement. All surfaces were machined to a mirror finish and gold plated. The first process was made easier by removing bearing B and the plate holding rods P from the stator assembly and collapsing the stator onto the tapered surfaces normally separating S1, S2, and S3. The assembly could then be honed as a unit.

The rotor is supported in a housing F1 by a pair of high-precision bearings G1 mounted in tandem and driven by means of a shaft Sh and pin drive Pn. The gold-plated spring K ensures good electrical contact between R and the shaft which, as well as carrying a pulley and a counter mechanism similar to that used for the self-centering cell described above, terminates in a mercury cup contact (the latter components are not shown in Figure 2). Shaft Sh is supported independently by bearings G2.

Housings F1 and F2 are filled with oil which is pumped via inlet I1 and outlet I2 through an external thermostat. This ensures both lubrication of the bearings G1 and G2 and temperature control of the cell. Two oil seals (one shown as O) are located at the top of F1 and bottom of F2.

The stator assembly is located central to the rotor assembled by means of a taper in the glass filled PTFE table T which in turn is located by three guide rods V (one shown). These serve as part of the mechanism for dismantling and assembly in that they ensure a coaxial withdrawal or insertion of the stator relative to the rotor. The rest of this mechanism which is not shown consists of a threaded drive rod mounted above the cell at the extremities of the rods W and connected when required, to the block A at points J (one shown) via a flexible coupling. The stator can thus be withdrawn for cleaning and filling the cell. During a slow lowering of the stator the test liquid is displaced upward through the shear gap and other gaps between S3, S2, and S1. It is important in design to ensure that these gaps are of such size as not to cause preferential flow in any channel system else air bubbles in the shear gap may result. At the top of stator component S1 a slot X serves as an overspill channel. An inserted syringe needle at Y is in line with the slot and excess liquid is aspirated via Y. The thermocouple Th also enters into the stator via a tube going right through this slot and into the body of S1 so that Th is located just under the surface of S1. A second thermocouple (not shown) is mounted in the housing F1.

Further specifications of the cell are given in Table 1. There is no provision for dynamic birefringence measurements.

This type of Couette cell has much better temperature control ($\pm 0.1\,°C$), and a wider temperature range than the self-centering stator cell described above. Static measurements can be obtained as well as measurements under shear rates of up to $\sim 10^4\,s^{-1}$, and the diversity of possible electrical measurement is an advantage, enabling even aqueous systems to be investigated with the microelectrode.

3. MACROMOLECULES REMAINING RIGID IN SHEAR

The expected flow-modified permittivity behavior of a spheroidal macromolecule with a dipole μ located along its symmetry axis was first considered by Saito and Kato.[4] Their model considers sufficiently dilute polymer solutions so that dipolar, hydrodynamic, or other interactions between particles are negligible and that the flow, electrical field $E = (E, 0, 0)$, and rotation axis are mutually orthogonal, the situation applicable to the types of flow described in Section 2. In the absence of macromolecules the flow velocity v^0 is given by

$$v^0 = G \cdot r, \qquad G = \begin{pmatrix} 0 & 0 & 0 \\ G & 0 & 0 \\ 0 & 0 & 0 \end{pmatrix} \tag{1}$$

where G is the shear tensor having as a component the single velocity gradient G applicable to the laminar flow in cylindrical geometry with narrow gaps. The angular momentum M of a particle is given by

$$\frac{dM}{dt} = L + \mu \wedge E \tag{2}$$

where L and $\mu \wedge E$ are moments of force due to viscous drag and electrostatic interaction. Saito and Kato use the analysis of Jeffery[12] for L. Solution of Eq. (2) provides the angular velocities in Eulerian angles θ, ϕ, and ψ and these are then involved in the Kirkwood[13] general diffusion equation for Brownian motion to provide a differential equation for the time-dependent angular distribution function of dipoles, $F(t, \theta, \phi, \psi)$. Both the electric vector E and shear rate G are assumed to vary sinusoidally with time. For solution the equation was expanded in terms of powers in $(\mu E_0/kT)$ and $(G_0 c_1/2kT)$ and associated Legendre polynomials. In these expressions

E_0 and G_0 are the zero-frequency intensity of the electric field and shear rate, respectively, and c_1 the friction constant of rotary motion of an ellipsoid about its minor axis. Complex recurrence relations are given for the coefficients in the series solution, and the influence of a time-dependent electric field on the viscoelasticity and electric polarization of such a rigid polar macromolecular solution is discussed. However, the relationships are exceedingly complex and as pointed out by Barisas[5] series convergence for the polarization is not achieved at high shear rates. Further, the requirement for G to be time dependent, that is for oscillatory flow, is an added complication that has not been applied to shear-modified permittivity investigations to date. With the simplification of constant G, spheroidal geometry, and in some cases also a time-independent electric field, Barisas has much simplified the treatment of Saito and Kato. The last simplification of a time-invariant electric field is of course not easily achieved practically; measurements are usually taken over a range of frequencies. When the restriction of time independence of field is theoretically applicable, the implication is that the theory applies to measurements taken at low frequency or extrapolated to zero frequency. The diffusion equation of Barisas is

$$\nabla^2 F - A_1 \frac{\partial F}{\partial \phi} - A_2 \frac{\partial F}{\partial \theta} + A_3 F = \frac{1}{D} \frac{\partial F}{\partial t} \tag{3}$$

where the coefficients are given by

$$A_1 = \gamma(1 + p \cos 2\phi) - \varepsilon \sin \phi / \sin \phi$$

$$A_2 = \gamma p \sin \theta \cos \theta \sin 2\phi + \varepsilon \cos \phi \cos \theta$$

and

$$A_3 = 3\gamma p \sin^2 \theta \cos 2\phi + 2\varepsilon \sin \theta \cos \phi$$

in which $\gamma = G/2D$, $\varepsilon = \mu E/kT$ and $p = (r^2 - 1)/(r^2 + 1)$ with D being the rotary diffusion coefficient of the particle about the minor axis and r the axial ratio. The solution for F can then be expanded as either

$$F = \sum_{ij} \gamma^i \varepsilon^j F_{ij} \tag{4}$$

as was done by Saito and Kato, or as

$$F = \sum_{ij} p^i \varepsilon^j \Phi_{ij} \tag{5}$$

in which F_{ij} or Φ_{ij} are written as sums of associated Legendre polynomials.

The coefficients are then established by recurrence relations derived from (3). The angular averaged polarization per particle is then obtained as

$$\langle\mu\rangle_G = \mu \int\int F \sin^2\theta \cos\phi \, d\theta \, d\phi \tag{6}$$

but this series of integrals does not converge for nonspherical particles if written in terms of F_{ij} when $\gamma > 2$ and thus there is an upper limit of G when expansion in terms of γ is invoked. Expansion using (5) does however give a converging solution and this is akin to Peterlin's treatment in the theory of dynamic birefringence.[14] Barisas[5] gives further details including the recurrence relations and a table of the changes in polarization $\langle\mu\rangle_G/\langle\mu\rangle_0$ as a function of $G/2D$ for ellipsoidal particles of various axial ratios, including $r = 1$ (spheres). Figure 3 shows this variation and it can be readily seen that the effect of shear on the low-frequency polarization is to reduce the component due to the rigid macromolecules, and to do so most dramatically over a fairly narrow range of G which depends upon the diffusion coefficient D. Shape has rather minor influence on the curves. Plots of polarization or the related permittivity changes as a function of shear rate G are termed Barisas plots in this review when rigid polar polymers for which the theory is applicable are the substrates.

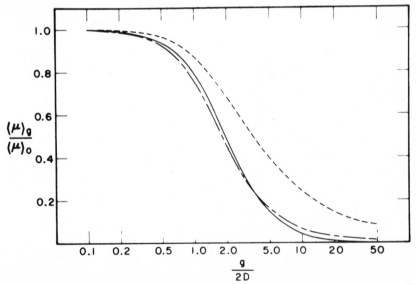

FIGURE 3. Predicted changes in the orientational polarization for rigid macromolecules undergoing shear induced rotation predicted by and reproduced from Barisas.[5] Data shown as polarization $\langle\mu\rangle_G$ at shear rate G relative to no flow conditions ($\langle\mu\rangle_0$) and at low electrical frequency. Results for spheres (———), rods (.....), and disks (––––).

Expansion in terms of γ and ε does give closed solutions for spherical particles and in this case also there need be no restriction on the frequency dependence of the field. The space-averaged complex polarization, $\langle \mu^* \rangle_{G,\omega}$ is given by

$$\langle \mu^* \rangle_{G,\omega} = \frac{\mu^2 E_0}{3kT} \left\{ \frac{[1 + (i\omega/2D)]}{[1 + (i\omega/2D)]^2 + (\gamma/2)^2} \right\} e^{i\omega t} \tag{7}$$

with a real component

$$\langle \mu \rangle_{G,\omega} = \frac{\mu^2 E_0}{3kT} \left\{ \frac{[1 + (\omega/2D)^2 + (G/4D)^2]}{[1 + (\omega/2D)^2 + (G/4D)^2 - 4(\omega/2D)^2(G/4D)^2]} \right\} \cos \omega t \tag{8}$$

Relation (8) shows that in the electric frequency plane at constant shear rate, the polarization due to polar spherical macromolecules may exhibit a maximum. This occurs at large shear rates ($G \geq 4D$) and at a frequency $\omega \sim G/2$. Although not stated by Barisas there is also a peak in the polarization in the shear plane at constant frequency if this is sufficiently high ($\omega \geq 1.547D$). If $\omega \gg 2D$ this peak is centered at $\sim G/2$ in the shear plane.

A number of polymers which fall into the category for which the Barisas theory is applicable have been studied by flow permittivity. Perhaps the most obvious group are the poly (α-amino acids) in the α-helical conformation. The flow-modified permittivity behavior of poly(γ-benzyl-L-glutamate) in helicogenic solvents has been investigated by a number of workers.[5,15,16] The polymer in helicogenic solvents has cylindrical symmetry, a high-persistence length, and a large permanent dipole along the major axis with zero components along the minor axes. At low frequency the dielectric contribution $\Delta\varepsilon_G'$ (relative permittivity of the solution above that of solvent) shows a typical reduction with shear rate and qualitatively matches very well the predictions of the Barisas theory. Figure 4 shows an example of PBLG of weight average molecular weight $= 1.49 \times 10^5$ dissolved in dioxane (concentration 1.08×10^{-3} g cm^{-3}). By curve fitting using the data of Barisas and needle geometry ($r = \infty$) the diffusion coefficient $D = 1.04 \times 10^4$ s^{-1} which is larger than estimated from dielectric spectroscopy.[15] Several factors may contribute to such a discrepancy. Polydispersity in molecular weight is not allowed for in the Barisas theory and it is expected that high polymer with its large dipole will dominate the flow-modified permittivity. The application of a theory developed for spheroidal geometry to cylindrical rods is also a source of discrepancy although it is unlikely that this would cause any major disparity between theory and experiment, particularly in view of the relative insensitivity of the theory

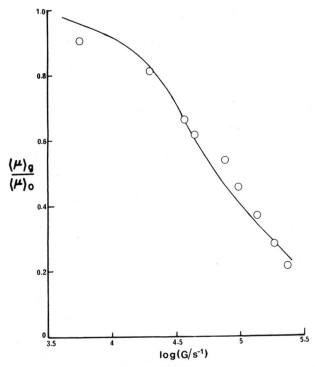

FIGURE 4. The experimentally observed variation ($\langle\mu\rangle_G/\langle\mu\rangle_0$) of polarization relative to zero shear for poly(γ-benzyl-L-glutamate) in dioxane. The curve is calculated on the basis of Barisas' theory[5] for needle geometry and a rotary diffusion coefficient of 1.04 $\times 10^4$ s^{-1} (after Block *et al.*[15]).

to axial ratio. As first pointed out by Barisas[5] the dependence of the variation in permittivity with shear is not very sensitive to shape, and shape estimates based on Barisas plots are quite tenuous. This is even more likely to be the case with polydisperse substrates.

Block *et al.*[15] have investigated the shear dependence of cell capacitance (simply related to permittivity) as a function of shear rate and frequency; the data are shown in Figure 5. This shows maxima which are predicted for spherical symmetry and obviously, and not unexpectedly, also occur for very asymmetric macromolecules although no theoretical framework is available in this case.

A more recent study[17] of poly(n-butyl isocyanate) and poly(n-hexyl isocyanates) show Barisas type of behavior in flow which is a consequence of the rigidity of this type of polymer. For poly(n-hexyl isocyanate), fractionated materials of low polydispersity were investigated as well as the

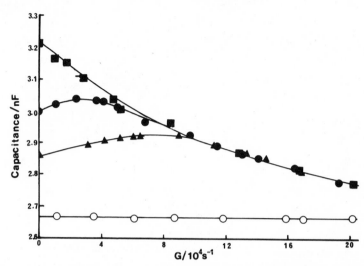

FIGURE 5. The variation of cell capacitance with shear rate for poly(γ-benzyl-L-gluta-mate) in dioxane at different measuring frequencies: ■, 400 Hz; ●, 1591 Hz; ▲, 4000 Hz. Pure solvent at 1591 Hz is shown by open circles (after Block et al.[15]).

polydisperse material. Fractions in general exhibited closer agreement between diffusion coefficients calculated from FMP vis-à-vis dielectric relaxation data than did the polydisperse material. An analysis of the effect of polydispersity on FMP was also presented and shown to largely account for the observed broadening of the change in $\Delta\varepsilon'(G)/\Delta\varepsilon'(0)$ with G.

For the polyisocyanates and poly(γ-benzyl-L-glutamate) most workers observe that at high shear rates the low-frequency dielectric contribution decays to zero.[†] This is not always the case with rigid macromolecules. For example, as illustrated in Figure 6 the system ethyl cellulose in benzene retains a component of $\langle\mu\rangle_G/\langle\mu\rangle_0$ at high shear rate, even for data extrapolated to zero concentration.[15] The cause has been ascribed to the presence of more than one component of dipole or more than a single polarization process. Thus, even if the macromolecule had a single summative dipole [such as is the case with α-helical poly(α-amino acids)], if that dipole were not colinear with a principle axis, then the two or even three resolved components would decay sequentially in a Barisas plot. Each decay would be governed by the appropriate diffusion coefficient. Other modes of relaxation such as internal modes or interfacial polarization, if faster than rotation, would also not contribute to the initial decay observed in the

[†] Takashima[16] has reported that a finite component remains at high shear rates.

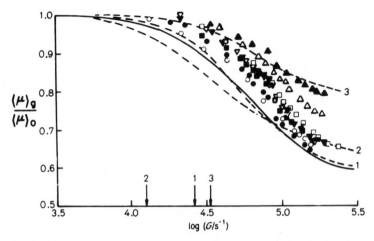

FIGURE 6. Flow-modified permittivity measurements on ethyl cellulose (number average molecular weight 4.5×10^4) in benzene at various concentrations (mg cm³): ○, 3.7; ●, 5,6; □, 7.05; ■, 7.75; ▽, 10.6; ▼, 10.95; △, 25.2; ▲, 50. The full line is data extrapolated to zero concentration and dashed curves refer to Barisas' theory[5] modified for retention of a nonshear relaxing component ($\langle\mu\rangle_\infty$); curve 1, $r = 1$, and $\langle\mu\rangle_\infty/\langle\mu\rangle_0 = 0.6$; curve 2, $r = \infty$ and $\langle\mu\rangle_\infty/\langle\mu\rangle_0 = 0.6$; curve 3, $r = 1$ and $\langle\mu\rangle_\infty/\langle\mu\rangle_0 = 0.8$. Marked intercepts correspond to $2D$ values on the G scale (after Block *et al.*[15]).

Barisas plots. These other processes still polarize the system at shear rates for which the slowest rotational diffusion process has relaxed out. The theory of Barisas can then be applied to the component which relaxes out *viz.* in terms of the ratio $(\langle\mu\rangle_G - \langle\mu\rangle_\infty)/(\langle\mu\rangle_0 - \langle\mu\rangle_\infty)$. Block *et al.*[15] show that flow-modified permittivity measurements are more sensitive than dielectric spectroscopy in resolving the principle reorientation process from other loss processes in this and other systems. For ethyl cellulose the data shown in Figure 6 provided the information that at least two loss processes were involved and that at infinite dilution only $\sim 40\%$ of the total polarization involved dipole reorientation of the major dipole. This component also is reduced by polymer association at finite concentrations while the remainder is not affected by the association.

Copolymers of γ-benzyl-D- and L-glutamates dissolved in *trans*-1,2-dichloroethylene containing 1 vol % N,N-formdimethylamide as deaggregant have recently been investigated[18] and show more than one component of polarization (Figure 7). This copolymer system is of interest because the inclusion of the D-isomer destabilizes the right-hand α helix of the pure L-homopolymer.[19] Dramatic increases in the component uninfluenced by shear can be seen and reflect the disruption of the helix and the develop-

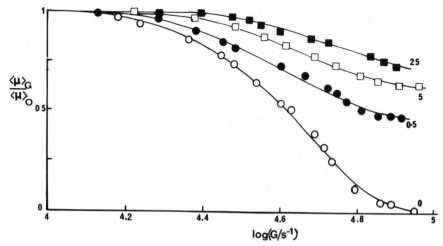

FIGURE 7. Barisas plots for copolymers of γ-benzyl-L- and D-glutamates in *trans*-1,2-dichloroethene with N,N-formdimethylamide (1 vol %) as deaggregant. Mole percentages of D-isomer indicated (after Tsangaris[18]).

ment of multiple dipolar relaxation mechanisms. Here again dielectric spectroscopy just shows a broader but still unresolved loss process with increasing D-isomer content. Diffusion coefficients as a function of enantiomorphic composition and molecular weight have been derived and are given in Table 2.

Preliminary observations have also been reported[11] of flow-modified permittivity of DNA samples in aqueous solution. This is a promising application of flow-modified permittivity since dielectric relaxation measurements on aqueous solutions present great difficulty particularly at low frequencies.

TABLE 2. Data[18] for the System Copoly (γ-benzyl-L- and D-glutamate) in *Trans*-1,2-dichloroethene with 1 vol % of N,N-formdimethylamide

Mol % D-isomer	10^4 Viscosity average molecular weight	Rotary diffusion coefficient/10^4 s^{-1}
0	17.0	1.39
0.5	14.2	1.44
2.0	11.5	1.56
5.0	10.9	1.62
8.0	8.4	1.67
25.0	3.6	2.06

4. FLEXIBLE POLYMERS IN SHEAR

In comparison to the flow-modified permittivity of nondeformable macromolecules the dielectric behavior of flexible macromolecules in flow is poorly understood. Peterlin and Reinhold have developed a model for the influence of a shear field on the average chain dimensions of a flexible coil and extended that model to a calculation of the changes in total dipole with shear.[20] The model is based on the Zimm–Rouse model and evaluates the in-flow average total dipole by vector summation of individual residue dipoles localized at each chain element. Marked differences in behavior are predicted for the case of such residue dipoles being parallel or perpendicular to the chain. At values of the number of perfectly elastic links Z sufficiently large to ensure a Gaussian distribution of such links, the predictions are that for perpendicular dipoles the orientation polarization per macromolecule $P_\perp$ is given in terms of the generalized flow parameter

$$\beta = M[\eta]\eta_0 G/RT$$

$$\text{by} \quad P_\perp = \frac{E\mu_0{}^2 Z}{3kT}\left[1 + \frac{\beta^2}{Z}\sum_{i=1}^{Z} F_i\right]$$

(9)

In these relations $[\eta]$ is the limiting viscosity number for the polymer of molecular mass M in a solvent whose viscosity is η_0 and F_i is a function of Z (given by the authors) whose sum, $\sum_i F_i = 0.4$ at very large Z.

For residue dipoles whose orientation is parallel to the chain Peterlin and Reinhold's model predict no change in polarization due to flow because the distortion of the chain leading to a larger total chain dipole is just balanced by the effect of chain orientation in the flow lines. Further the authors stipulate that their theory is only applicable at low values of β since the Rouse model is deficient at significant chain distortions. However, a saturation value at high shear rate is predicted.

Experimentally, studies of flow-modified permittivity on flexible polymers are not in satisfactory agreement with Peterlin and Reinhold's theory. Although early work on the flow-modified permittivity of flexible macromolecules is sometimes contradictory[21] (probably in part due to experimental artefacts) the results of Wendisch[22] and ourselves[23] support the prediction of shear-induced permittivity increases for polymers having perpendicular chain dipoles. However the magnitudes are larger than predicted. Rheinhold and Peterlin[20] discuss the data of Wendisch,[22] who studied nitrocellulose (both parallel and perpendicular components) in *n*-butylacetate and comment that for the increase in permittivity observed β values some 10^3 times those used would have to be invoked to match theory with experiment. They suggest that either intermolecular effects or errors in molecular weight are responsible, but it is difficult to see how a sufficiently large error in the latter could have occurred.

Block *et al.*[23] have investigated the polymer systems poly(styrene), poly(*p*-ethoxystyrene), poly(methyl methacrylate) and poly(*N*-vinylcarbazole), all polymers of the perpendicular dipole only class. Figure 8 shows data obtained for the last two polymers. These, particularly poly(*N*-vinylcarbazole), show the onset of saturation.

Comparative experiments on poly(styrene) and poly(*p*-ethoxystyrene) strongly suggested that the observed increases in relative permittivity ($\delta\varepsilon_G'$) due to shear were only marginally, if at all, dependent on the residue dipole

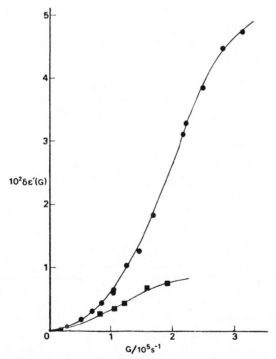

FIGURE 8. Flow-modified permittivity in solutions of poly(methyl methacrylate) (PMMA) and poly(*N*-vinyl carbazole) (PNVC) in benzene. $\delta\varepsilon'(G)$ = relative permittivity of solution at shear rate G less the relative permittivity of still solution. ●, Data for PMMA (number average molecular weight 1.04×10^5 and concentration 99 kg m^{-3}; ■, data for PNVC (number average molecular weight 1.25×10^6 and concentration 50 kg m^{-3}) (after Block *et al.*[23]).

although the dipole for poly(p-ethoxystyrene) is ~ 4 times that for styrene. However solvent polarity had a marked influence on the magnitude of the effect. Empirically the data on a number of polymers of different molecular weights, obtained in different solvents and at various concentrations c obeyed the relationship

$$\Delta \varepsilon_G' \propto \frac{(\varepsilon_0' + 2)\eta_0^2(\eta_r - 1)^2 G^2}{c\langle p_n \rangle^{1.19}} \tag{10}$$

where ε_0' is the solvent relative permittivity, η_r is the relative viscosity, and $\langle p_n \rangle$ the number average degree of polymerization. Relation (9) can be expressed in terms of β if the identity $[\eta] = (\eta_r - 1)/c$ at finite concentrations c is made, but then the concentration, molecular weight dependence, and most strikingly, the residue dipole dependence do not fit Eq. (9). Further, solvent relative permittivity ε_0' participates in terms of what appears as a Clausius-Mosotti internal field term $(\varepsilon_0' + 2)$.

The large discrepancies between the predictions of the Peterlin and Reinhold theory and experimental observation suggest either some fundamental error in the model, or that effects other than changes of polarization due to changes in residue dipole orientation contribute to changes of permittivity with shear. It should be noted that the theories of Peterlin and Reinhold and that of Barisas do not have a common basis in that the former considers that the influence of flow is to orientate and deform a flexible macromolecule so as to progressively align it in the flow direction, while for the latter the molecule is not deformed but rotated by the flow. A combined motion of rotation and local chain reptation probably occurs for flexible polymers, and their ability to polarize in the field must be a time-dependent as well as a spatial property. Also, the Peterlin and Reinhold theory does not hold at large deformations nor at polymer concentrations such that chain entanglements become significant. Unfortunately, flexible polymers possess much lower coil dipoles than the more rigid rodlike structures and this has the effect of reducing their contribution $d\varepsilon'/dc$ to the permittivity of the system. Higher concentrations of flexible polymers of higher molecular weights than is the case for rigid polar macromolecules are required for the measurement of effects. There are also possibilities for other factors being responsible for the observed magnitudes of permittivity increases in flow. Changes in high-frequency polarizability related to the changes causing flow birefringence are present but estimates suggest that these are too small to explain the observed permittivity increases.[23] A change in solvent packing in the coil on deformation could lead to a high number density of dipoles in the media, but as estimated by Block et al.[23]

this effect of shear striction is not likely to be of sufficient magnitude. Coil deformation would be expected to influence the magnitude of the internal field correction and this would thus be shear rate dependent. Although no quantitative theory exists for such an effect, it could be significant, since not only the residue dipoles of the chain but also solvent occluded by, and even near a polymer coil would experience field alterations. A similar effect is well known to be of importance as the form birefringence term in flow birefringence. Certainly the sensitivity of flow-modified permittivity to solvent nature, when the substrate is a flexible polymer, lends support to the view that variation in the internal field term may be a dominant factor. At present the phenomenon in systems of flexible macromolecules still requires extensive experimental and theoretical investigation.

5. COLLOID SYSTEMS IN SHEAR

In an endeavor to study the importance or otherwise of the internal field problem discussed in Section 4 we investigated a number of sterically stabilized organic colloids by the technique of flow-modified permittivity.[24] We observed an at first unexpected phenomenon with these systems as shown in Figure 9. Forced resonances occur showing peaks in the real part of the permittivity at defined shear rates G_m, which is proportional to the electric measuring frequency f as shown in Figure 10. We believe the cause for the forced resonance is due to a coupling between the electric and shear fields and can occur whenever a particle with its own internal time-dependent dielectric polarization is spun in flow. For the phenomena discussed in this and the next section there is no requirement that the dielectric properties internal to the particles be altered in any way by the flow, only that the orientation of the time-dependent particle polarization is. The observations reported in this section refer to cross-linked poly(ethyl acrylate) particles stabilized with a "comb" graft copolymer formed from poly-(12-hydroxy-stearic acid) terminated by glycidylmethacrylate and methyl methacrylate (1:1) and dispersed in hydrocarbons such as heptane. However it is our view that the phenomenon is probably not restricted to these systems (see Section 6). What is necessary for forced resonance to occur is that the polarization of the dispersed particles be time dependent and that their dielectric relaxation frequency be of similar order to the shear rate. The organic colloids studied have been shown by dielectric spectroscopy to possess low- ($\sim 10^{-1}$–10^3 Hz) and high- ($\sim 10^8$–10^{10}) frequency processes,[24] and it is the former which is responsible for the resonance phenomenon.

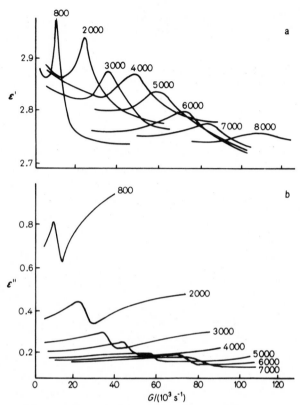

FIGURE 9. Forced resonances exhibited by an organic colloid in flow-modified permittivity. Numbers refer to the electrical frequency/Hz (after Block *et al.*[24]). (a) Relative permittivity (ε'), (b) loss (ε'').

A very simplified model for the resonance can readily be developed as follows. We assume that the particles are spherical,[†] that the relaxation process involved is of the Debye–Pellat form having only a single relaxation time τ, that the torque developed between the electric field and induced polarization does not alter the angular rotational frequency ν of the particle in the shear field, and that the disordering due to Brownian rotational diffusion is negligible. Since at appropriate electrical frequencies ω the polarization vector $\mathbf{P}(t)$ time lags the local field vector $\mathbf{E}(t)$, and since the particle is spinning due to shear, $\mathbf{P}(t)$ and $\mathbf{E}(t)$ are no longer colinear. Thus if the shear field is such that the particle rotates about the $\mathbf{k}$ axis and $\mathbf{E}(t)$ is along the $\mathbf{i}$ axis, then there will be components of $\mathbf{P}(t)$ along the $\mathbf{i}$ and

[†] This is the case for the organic colloids studied.

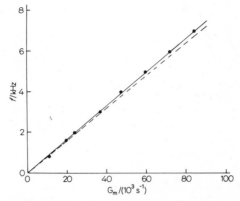

FIGURE 10. The dependence of the shear rate maxima (G_{max}) of Figure 9 on frequency f. The dashed line has the theoretical slope of $1/4\pi$ (after Block *et al.*[24]). The slight discrepancy is attributed to a small error in measured gap dimension used to calculate G.

j axes. If $\mathbf{P}(t)$ and $\mathbf{E}(t)$ are mutually inclined at θ at any time t,

$$\frac{d\mathbf{P}(t)}{dt} = \frac{\partial \mathbf{P}(t)}{\partial t} + \frac{\partial \mathbf{P}(t)}{\partial \theta} \frac{d\theta}{dt} \tag{11}$$

where the first term on the right-hand side of (11) refers to the time-dependent polarization at fixed θ and the second term to the changes of polarization due to rotation of the particle alone. For the Debye–Pellat process

$$\tau \frac{\partial \mathbf{P}(t)}{\partial t} + \mathbf{P}(t) = \epsilon_0 \Delta \varepsilon_1' \mathbf{E}(t) \tag{12}$$

where $\Delta \varepsilon_1'$ is the change in relative permittivity over the Debye–Pellat processes and ϵ_0 is the permittivity of free space. Further, $d\theta/dt = \nu = G/2$ for the applied simplifying conditions and Couette flow.[12]

From this relation, Eq. (12) and the angular variation of $\mathbf{P}(t)$, expression (11) can be written in terms of the time-dependent components P_i and P_j. For $E(t) = gE_0 e^{i\omega t}$, where g is the internal field correction

$$\tau \frac{dP_i}{dt} + P_i + \nu\tau P_j = g\epsilon_0 \Delta \varepsilon_1' E_0 e^{i\omega t} \tag{13a}$$

$$\tau \frac{dP_j}{dt} + P_j - \nu\tau P_i = 0 \tag{13b}$$

These simultaneous differential equations are readily converted to second-order differential equations in either P_i or P_j. For the former

$$\tau^2 \frac{d^2 P_i}{dt^2} + 2\tau \frac{dP_i}{dt} + (1 + \nu^2\tau^2)P_i = (1 + i\omega\tau)g\epsilon_0 \Delta\epsilon_1' E_0 e^{i\omega t} \quad (14)$$

which has the solution

$$P_i = C_1 e^{(i\nu - 1/\tau)t} + C_2 e^{-(i\nu + 1/\tau)t}$$

$$+ \frac{1 + i\omega\tau}{\{1 + \nu^2\tau^2 - \omega^2\tau^2 + 2i\omega\tau\}} g\epsilon_0 \Delta\epsilon_1' E_0 e^{i\omega\tau} \quad (15)$$

in which C_1 and C_2 are integration constants. Only the long-time, non-decaying solutions of (15) are of interest in relation to the effect observed. Relation (15) is readily converted to give the relative permittivity and loss of an assembly of particles under shear. Thus for a Clausius–Mosotti internal field and with a volume fraction of particles ϕ in solvent of relative permittivity ϵ_0' and particle low frequency relative permittivity ϵ_1'

$$\epsilon'(\nu, \omega) - \epsilon_0' = \phi \left\{ \frac{\epsilon_0' + 2}{\epsilon_1' + 2} \right\}^2 \Delta\epsilon_1' \left\{ \frac{1 + \Gamma}{(1 + \Gamma)^2 + \omega^2\tau^2(1 - \Gamma)^2} \right\} \quad (16)$$

$$\epsilon''(\nu, \omega) = \phi \left\{ \frac{\epsilon_0' + 2}{\epsilon_1' + 2} \right\}^2 \Delta\epsilon_1' \left\{ \frac{\omega\tau(1 - \Gamma)}{(1 + \Gamma)^2 + \omega^2\tau^2(1 - \Gamma)^2} \right\} \quad (17)$$

where $\Gamma = \nu^2\tau^2/(1 + \omega^2\tau^2)$ and ϵ', ϵ'' are the relative permittivity and loss of the dispersion in shear. The last, bracketed terms in (16) and (17), determine the shear and frequency dependence of ϵ' and ϵ''. From (16) it can be predicted that a peak will appear in the permittivity–shear dependence provided $\omega \geq 1/\sqrt{3}\tau$ and that when $\omega \gg 1/\tau$ this peak will have a maximum located at $\nu_{max} = \omega$. Since $\nu = G/2$ and $\omega = 2\pi f$ this condition occurs at $G_{max} = 4\pi f$. Thus the observed proportionality between f and G_{max} is predicted as shown in Figure 10. Further, the form of the frequency and shear rate dependence of both the relative permittivity and loss reflect the generally observed behavior as a comparison of Figures 9 and 11 shows. However, relation (16) predicts a halving of the resonance peaks from a condition of low ν_{max} to one of high ν_{max} but the experimental indications are that a greater relative decrease in intensity occurs. This may be due to simplification in the theory. Recently a more sophisticated analysis in both two and three dimensions of the situation has been given in which the influence of Brownian motion on nonspherical but spheroidal particles undergoing a uniform flow-induced rotation are considered.[25] Further

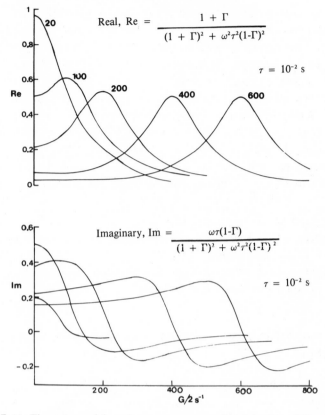

$$\text{Real, Re} = \frac{1 + \Gamma}{(1 + \Gamma)^2 + \omega^2\tau^2(1-\Gamma)^2}$$

$$\tau = 10^{-2} \text{ s}$$

$$\text{Imaginary, Im} = \frac{\omega\tau(1-\Gamma)}{(1 + \Gamma)^2 + \omega^2\tau^2(1-\Gamma)^2}$$

$$\tau = 10^{-2} \text{ s}$$

$G/2 \text{ s}^{-1}$

FIGURE 11. The nature of forced resonances as predicted by Eqs. (16) and (17).

complicating factors such as the influence of a distribution of particle sizes (effecting the Brownian motion), a distribution of relaxation times, non-uniform precessive motion in uniform flow fields for nonspherical particles, and the influence on particle rotation rates of torque between noncolinear **P** and **E** await quantitative theoretical evaluation. It may well be that some of these factors are the cause of the apparently more extensive attenuation of resonances with shear rate. The induced torques due to interactions **P** and **E** have importance in the rheology of such systems as discussed further in Section 6.

The phenomenon described above may have an application in the study of polyelectrolytes. Experimental observation has shown that the shearing of a liquid media having bulk conductance does not result in observable forced resonance peaks. This is because the phenomenon as envisaged requires a localized polarization which is associated with a region of space whose axes are rotated by the shear field. A polyelectrolyte coil with its associated ion atmosphere is known to have such low-frequency localized losses, but since intensity measurements in the frequency plane without shear are heavily screened by bulk conductance their observation by conventional dielectric means is very difficult. Flow-modified permittivity measurements, especially after the advent of more rigorous theory, may well prove a practical alternative for such studies.

It is possible that the observations alluded to earlier in relation to flexible polymers may also have their origin in such a localized coil polarization interacting with rotation in the shear field. If this were so then the data shown in Figure 8, for the poly(styrene) and poly(p-ethoxystyrene) systems[23] would reflect the onset of forced resonance peaks. However, it should be stressed that in these cases the condition that $2\pi f \geq 1/\sqrt{3}\tau$ does not seem to hold since there are no dielectric loss processes observable in the appropriate range. Also the condition $G_{max} = 4\pi f$ would place peaks at much lower shear rates. Only a marked failure of the simple theory given above could reconcile these problems.

6. ELECTROVISCOUS PHENOMENA AND THE WINSLOW EFFECT

In 1947 W. M. Winslow[6] reported that dispersions of a range of substances such as starch, limestone, gypsum, flour, gelatine, carbon, etc. in a variety of nonconducting liquids including mineral oils showed greatly enhanced viscosities when subjected to electric fields. Winslow patented a number of devices based on this electroviscous effect. In a subsequent publication[7] he drew attention to the fact that the effect is particularly good if the particles are of a semiconducting nature and have a high dielectric constant. The explanation advanced by Winslow for this form of electroviscous effect is based on polarization of the particles leading to aggregation and a tendency for linear fibril formation across the direction of flow. Indeed in static situations such fibration is well established.

This type of electroviscous or Winslow effect has been investigated by a number of researchers,[7,26–30] who have not been unanimous in ascribing the above mechanism. In particular, Klass and Martinek[26,27] in studies of silica or calcium titanate dispersed in a vehicle described as "naphthenic" suggested that interfacial polarization causes an increase in viscosity because of consequent interactive forces between polarized particles. They point out that in the time scale of the effect, aggregation is unlikely and is

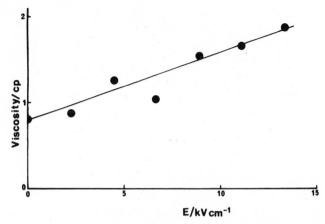

FIGURE 12. Electroviscosity in a sample of organic colloid dispersed in heptane (20 wt %). Mean particle diameter 0.1 μm (after Gregson[31]).

indeed unnecessary. Uejima[28] ascribes the effect to surface water films and their polarization and Okagawa et al.[29,30] have developed a theoretical model based on the rotational behavior of ellipsoidal dielectric particles in electric fields which predicts a form of electroviscosity.

Recently we have observed that the organic colloids whose flow-modified permittivity behavior is described in Section 5 exhibit enhancement of viscosity under field. Figure 12 illustrates this behavior for a 20% volume fraction dispersion in heptane and much greater increases in the effect occur at higher concentrations.[31] Indeed we have inadvertently stalled the 0.185-kW motor driving the equipment by the application of quite moderate dc voltages.

We believe that the mechanism for the effect in our systems and probably for the Winslow effect observed in other dispersions is at least in part due to energy dissipation consequent upon polarization and rotation of the particles. Put another way, the field **E** provides a torque on the particle because of the presence of the polarization component P_j whose forced rotation under flow thereby provides a work term. This concept is somewhat different from that proposed by Okagawa et al.[29,30] in that these authors do not consider that a time-dependent polarization of the particle is involved. Rather they ascribe the effect to a permanent polarization (that is, electret character), an induced polarization, or both. As they point out[30] for spherical particles lacking a permanent dipole, an induced polarization instantly following the field cannot result in an electrical torque. Thus there is no effect for spherical particles as distinct from asymmetric particles, unless a permanent dipole is assumed. The model of Okagawa *et al.* thus does not explain a Winslow effect for spherical particles unless they carry a permanent polarization, and thus differs from our concept of the time dependence of polarization being responsible. Also, our mechanism for an electroviscous effect would be operative even at high dilutions and although a particle-interactive mechanism of the type envisaged by Klass and Martinek may well supplement the phenomenon at higher concentrations, some viscous drag must be induced by the noncolinear nature of the polarization and field. Certain of the observations by Klass and Martinek[26] lend support to the time-dependent polarization model. For their silica dispersions, the electroviscosity decreases rapidly with shear rate, and since their silica system has a dielectric loss at $\sim$400 Hz this observation is consistent with the model of spinning particles whose polarization is increasingly attenuated with increased shear rate. The electroviscosity at constant shear rate also decreases with increasing electrical frequency f when $f \gtrsim 200$ Hz and approaches the no-field value at $f > 10^5$ Hz (Figure 13). It is noteworthy that a marked discontinuity in their curve, reminiscent, if inverse, of the forced resonances shown in Figure 9, occurs at $f \sim G/\pi$ for which Klass and Martinek offer no explanation. However, further confirmation of the cause or causes for the Winslow effect still await both a quantitative theory and more experimental data.

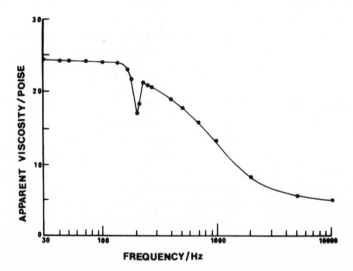

FIGURE 13. Electroviscosity in a silica dispersion (0.46 vol fraction) after Klass and Martinek.[26] The dependence of the phenomenon on the frequency of a 9.3 kV cm^{-1} field and at a constant shear rate of 753 s^{-1} is shown. Viscosity at zero field is 2.2 P.

There are a number of potential applications for electroviscous effects particularly since very considerable increases in viscosity can be achieved. Winslow[6] in his original patents cites the use in transmission coupling between two disks or disk and cup with an electroviscous fluid between them; he also describes an electrofluid relay based on the same effect. Thus both clutch and brakes based on the Winslow effect have been constructed.[6,7] Boeing[32] has developed a hydraulic vibrator based on electric-field valves worked by Winslow effect and a clutch-driven loudspeaker has also been described.[7] Further research into both the applications and, in particular, the fundamentals of the effect, are in the author's view very desirable, particularly since it is probable that the optimum materials for the phenomenon have not yet been realized. A low-frequency loss process in the dispersed particles seems to be a necessary prerequisite for the phenomenon but this need not necessarily be of an interfacial nature. A combination of the steric stabilization of organic colloids in order to provide stable dispersions in oils plus a "lossy" bulk phase for the colloid material promises to be a winning combination.

7. POSSIBLE FUTURE DEVELOPMENTS

In the preceding pages indications have been given of the use of the technique of flow-modified permittivity in the study of macromolecules and of the phenomenon of electroviscosity of the type described by Winslow. However, our knowledge of these effects is at present still in its infancy. Hopefully future work will strengthen our understanding of these effects and render them powerful techniques in the characterization of polymers in flow and provide useful devices based on the Winslow effect. At this time a number of developments are particularly urgent, or are likely to lead to important extensions of the use of flow-modified permittivity. In our view these important developments comprise the following:

(1) There is a need for further theoretical work in the behavior of flexible polar macromolecules in shear so that a clearer quantitative interpretation of permittivity changes in flow can be made.

(2) Similarly the theory of the forced resonance in spinning particulates requires refinement, particularly to take into account torque effects on angular velocity.

(3) The Winslow effect requires much more theoretical and practical investigation. As regards the former, any model under (2) will

lead to a quantifiable viscous term which should apply in the limit of infinite particle dilution.

(4) The phenomenon of forced resonance should have a counterpart in macromolecular coils having a low-frequency local polarization. This could be particularly important in the study of polyelectrolytes and biopolymers in their natural environment since bulk conductance appears invariant under shear.

It is the hope and belief of this author that, with the developments indicated above, the phenomena of flow-modified permittivity and electroviscosity will become powerful techniques in our studies and uses of macromolecules in flow.

REFERENCES

1. A. Peterlin, W. Heller, and M. Nakagi, *J. Chem. Phys.* **28**, 470–476 (1958); A. F. Stevenson and H. L. Bhatnagar, *J. Chem. Phys.* **29**, 1336–1339 (1958); M. Nakagi and W. Heller, *J. Polym. Sci.* **38**, 117–131 (1959); W. Heller, E. Wada, and L. A. Papazian, *J. Polym. Sci.* **47**, 481–484 (1960); A. Peterlin and C. Reinhold, *J. Chem. Phys.* **40**, 1029–1032 (1964); **42**, 2172–2176 (1964); C. Reinhold and A. Peterlin, *Physica* **31**, 522–540 (1965); A. F. Stevenson, *J. Chem. Phys.* **49**, 4545–4550 (1968); R. Tabian, M. Nakagi, and L. Papazian, *J. Chem. Phys.* **52**, 4294–4305 (1970).
2. L. de Brouckere and M. Mandel, *Adv. Chem. Phys.* **1**, 77–118 (1958); W. H. Stockmayer, *Pure Appl. Chem.* **15**, 539–543 (1967); H. Block and A. M. North, *Adv. Mol. Relaxation Proc.* **1**, 309–374 (1970); A. M. North, *Chem. Soc. Rev.* **1**, 49–72 (1972); H. Block, *Adv. Polym. Sci.* **33**, 93–167 (1979).
3. M. W. Volkenstein, *Configurational Statistics of Polymeric Chains*, Interscience, New York (1966); P. J. Flory, *Statistical Mechanics of Chain Molecules*, Interscience, New York (1969); P. J. Flory, *Pure Appl. Chem.* **26**, 309–326 (1971); J. E. Mark, *Acc. Chem. Res.* **7**, 218–225 (1974).
4. N. Saito and T. Kato, *J. Phys. Soc. Japan* **12**, 1393–1402 (1957).
5. B. G. Barisas, *Macromolecules* **7**, 930–933 (1974).
6. W. M. Winslow, U. S. Patent 2, 417, 850 (1947).
7. W. M. Winslow, *J. Appl. Phys.* **20**, 1137–1140 (1949).
8. G. I. Taylor, *Proc. R. Soc. London Ser. A* **157**, 546–564 (1936).
9. J. V. Champion, *Proc. Phys. Soc.* **75**, 421–433 (1960).
10. H. Block, E. M. Gregson, W. D. Ions, G. Powell, R. P. Singh, and S. M. Walker, *J. Phys. E: Sci. Instrum.* **11**, 251–255 (1978).
11. H. Block, E. M. Gregson, A. Qin, G. Tsangaris, and S. M. Walker, *J. Phys. E: Sci. Instrum.* **16**, 896–902 (1983).
12. G. B. Jeffery, *Proc. R. Soc. London* Ser. A **102**, 161–179 (1923).
13. J. G. Kirkwood, *Rec. Trav. Chim.* **68**, 649–660 (1949).
14. A. Peterlin, *Z. Phys.* **111**, 232–263 (1938).

15. H. Block, W. D. Ions, G. Powell, R. P. Singh, and S. M. Walker, *Proc. R. Soc. London* Ser. *A* **352**, 153–167 (1976).
16. S. Takashima, *J. Phys. Chem.* **74**, 446–452 (1970).
17. H. Block, E. M. Gregson, A. Ritchie, and S. M. Walker, *Polymer* **24**, 859–864 (1983).
18. G. Tsangaris, Ph. D. thesis, University of Liverpool (1981).
19. H. Block, *Poly(γ-benzyl-L-glutamate) and other glutamic acid containing polymers*, Vol. 9, *Polymer Monographs* (M. B. Huglin, ed.) Gordon and Breach Science Publishers, New York (1983).
20. A. Peterlin and C. Reinhold, *Kolloid Z. Z. Polym.* **204**, 23–28 (1965).
21. B. L. Funt and S. G. Mason, *Can. J. Chem.* **29**, 848–856 (1951); H. Hartmann and R. Jaenicke, *Z. Phys. Chem. N. F.* **6**, 220–241 (1956).
22. P. Wendisch, *Kolloid Z. Z. Polym.* **199**, 27–31 (1964).
23. H. Block, W. D. Ions, and S. M. Walker, *J. Polym. Sci. Polym. Phys. Edn.* **16**, 989–998 (1978).
24. H. Block, K. M. W. Goodwin, E. M. Gregson, and S. M. Walker, *Nature* **275**, 632–634 (1978).
25. H. Block, E. Kluk, J. McConnell, and B. K. P. Scaife, *J. Colloid Interface Sci.* **101**, 320–329 (1984).
26. D. L. Klass and T. W. Martinek, *J. Appl. Phys.* **38**, 64–74 (1967).
27. D. L. Klass and T. W. Martinek, *J. Appl. Phys.* **38**, 75–80 (1967).
28. H. Uejima, *Jpn. J. Appl. Phys.* **11**, 319–325 (1972).
29. A. Okagawa, R. G. Cox, and S. G. Mason, *J. Colloid Interface Sci.* **47**, 536–567 (1974).
30. A. Okagawa and S. G. Mason, *J. Colloid Interface Sci.* **47**, 568–587 (1974).
31. E. M. Gregson, Ph. D. thesis, University of Liverpool (1979).
32. H. T. Strandrud, *Hydraulics and Pneumatics*, 139–143 (1966).

Entanglement Effects in Polymer Solutions

WILLIAM W. GRAESSLEY

1. INTRODUCTION

Viscoelastic behavior in polymer solutions is a reflection of dynamical processes at the molecular level. These processes are driven by Brownian motion, and they govern diffusion and the rate of rearrangement of chain orientation and conformation. Mechanical response and molecular relaxation go hand in hand. Properties which depend on large-scale chain motions are strong functions of the molecular architecture (chain length, long chain branching, etc.) at all levels of polymer concentration.

At high concentrations the response acquires some special characteristics which are attributed to chain entanglement. Entanglement effects are common to all polymer species and appear to derive simply from the topological constraint of molecular uncrossability: each chain must move in such a way as never to pass through the backbones of other chains. Beyond a certain chain length, dependent on both species and concentration, the uncrossability constraint slows the rate of large-scale conformational rearrangement, but leaves the local chain motions essentially unaffected. Entanglement interactions dominate all properties which depend primarily on large-scale chain motions. They increase in importance with chain length

WILLIAM W. GRAESSLEY ● Exxon Research and Engineering Co., Corporate Research—Science Laboratories, Annandale, New Jersey 08801.

and concentration, and they appear to yield a set of quite general laws relating chain architecture and properties.

The object of this chapter is to summarize the experimental observations for flexible linear and branched chains at high concentrations and to describe recent theories about the entanglement effect. Linear viscoelastic behavior is emphasized, although a few specific nonlinear properties are also considered. Linear response lends itself most readily to molecular interpretations because it reflects only the unperturbed dynamics of the system. Nonlinear response depends also on the effect of a finite displacement from equilibrium. The behavior of structurally uniform systems (model polymers) is the primary focus; the complicated and still poorly understood effects of molecular weight distribution will not be discussed. The intermediate concentration regime is also omitted. The effect of entanglements is less settled there, and a separate chapter would be required for adequate coverage. Related material is discussed elsewhere in this volume, i.e., chain conformation at equilibrium in concentrated solutions (Chapter 4), molecular models and chain dynamics in dilute solutions (Chapter 1), and the fast dynamics of chains in concentrated solutions (Chapter 5). The books of Ferry[1] and de Gennes[2] and reviews by Berry and Fox[3] and Graessley[4,5] provide more detailed discussions. The bibliography is by no means exhaustive. Recent work is emphasized, but the references in those papers lead back to the original contributions.

2. CHAIN ARRANGEMENT AND MOTION IN CONCENTRATED SOLUTIONS

From results obtained by neutron scattering there is now a reasonably good understanding about equilibrium chain conformation in concentrated polymer liquids, at least on distance scales which are large compared to the monomer size (see Chapter 4, for example). The pervaded volumes of the chains interpenetrate extensively, and the intramolecular excluded volume interactions are effectively screened. The chains assume their theta-solvent dimensions. They obey random walk statistics, and the distribution of large-scale conformations is governed only by molecular architecture and C_∞, the characteristic ratio of the species.[6]

Given these features, an undiluted polymer or concentrated solution probably resembles a kind of random meshwork of chain contours, each chain following some tortuous (random walk) trajectory through a tangled mass of surrounding chains (Figure 1). Locally, i.e., at the monomer scale,

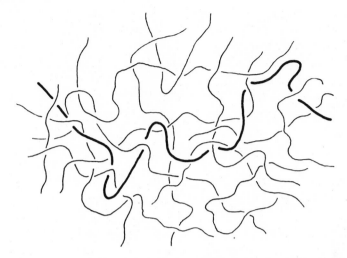

FIGURE 1. A polymer chain and surroundings in an entangled liquid.

the physical structure and properties are almost certainly independent of molecular architecture. This picture should apply over some range of concentrations, the spacings simply growing larger with dilution until a regime of sufficiently weakened screening is reached, the semidilute region,[2] where chain dimensions then begin to increase. That transition is probably a rather gradual one; estimates have placed it in the range of 10%–25% polymer.[7]

The dynamics of chains at high concentration are rather more complicated. As in dilute solution, the elementary motions of the chains are probably cooperative bond rotations, with precise details and rates that are highly individual to each species and local environment.[8] However they occur, their cumulative effect with time is to relax the correlation with past conformations over progressively larger distances along the chain. Driven by these rapid elementary motions, each molecule gradually drifts from one large-scale conformation to another. Like excluded volume, the hydrodynamic interactions between units on the same chain are well screened by the other chains.

Our main concern here is with the properties that depend on the large-scale chain motions, i.e., the slow dynamics. The elementary motions are still important in setting the time scale, but precise details tend to be averaged out as the relaxation propagates to larger chain distances. The effect of local motion on the slow dynamics can be represented by an empirical parameter, the apparent monomeric mobility, or its reciprocal, the monomeric friction coefficient.[1] A concentrated polymer solution is assumed to behave locally as a Newtonian fluid. Each molecule moves like the corresponding Rouse chain (see Chapter 1) would move if subjected to the constraint that it cannot cross the contours of the other chains. Thus, stripping the problem to its essentials, the basic picture for the slow dynamics in concentrated solutions is that of a highly overlapping collection of uncrossable Rouse chains. Each chain moves randomly in a locally viscous medium which is permeated by a mesh of similarly moving, uncrossable contours. The local motions are hardly affected by uncrossability since they involve only short chain distances. The large-scale motions are strongly affected because the chains must somehow diffuse around one another in order to take up entirely new conformations.

This preliminary discussion of entanglement effects is based almost entirely on inferences drawn from linear viscoelastic behavior and comparisons with Rouse model predictions for the fast dynamics of long chains with the slow dynamics of chains which are below the entanglement threshold. The following section treats linear viscoelasticity in the context of concentrated polymer solutions.

3. LINEAR VISCOELASTICITY IN CONCENTRATED SOLUTIONS

It is a remarkable fact, well substantiated experimentally, that all information about the linear mechanical response of any liquid to volume-conserving deformations can be expressed in terms of a single time-dependent function.[1] There are a number of equivalent functions, one being the shear stress relaxation modulus $G(t)$. That property describes the time dependence of shear stress $\sigma(t)$ following a small jump in shear strain γ_0 (Figure 2). If γ_0 is small enough, the stress at any later time is directly proportional to strain:

$$\sigma(t) = G(t)\gamma_0 \tag{1}$$

The initial modulus is $G(0)$, and $G(\infty) = 0$ for any liquid. Stress and strain for any history of sufficiently small or slow shear deformations are related

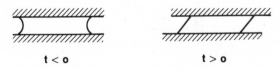

t < o   t > o

FIGURE 2. A simple shear step strain.

through $G(t)$ by the Boltzmann superposition formula. For a simple shear history $\gamma(t)$,

$$\sigma(t) = \int_{-\infty}^{t} G(t - s) \frac{d\gamma(s)}{ds} \, ds \tag{2}$$

At the molecular level this means that the disturbance from equilibrium is always so small that the responses to past strain impulses are additive.

Storage and loss components of the complex dynamic modulus, $G^*(\omega) = G'(\omega) + iG''(\omega)$, are the measured quantities at steady state in the commonly used oscillatory shear experiment. These are connected with $G(t)$ through Eq. (2) $[\gamma(t) = \gamma_0 \exp(i\omega t)]$:

$$G^*(\omega) = i\omega \int_{0}^{\infty} G(t) \exp(-i\omega t) \, dt \tag{3}$$

where

$$G(0) = G'(0) = \frac{2}{\pi} \int_{0}^{\infty} G''(\omega) \, d \ln \omega \tag{4}$$

Similarly, for the steady state response to a small and constant shear rate $\dot{\gamma}$,

$$\eta_0 = \int_{0}^{\infty} G(t) \, dt = \lim_{\omega \to 0} \frac{G''(\omega)}{\omega} \tag{5}$$

$$J_e^0 = \frac{1}{\eta_0^2} \int_{0}^{\infty} tG(t) \, dt = \frac{1}{\eta_0^2} \lim_{\omega \to 0} \frac{G'(\omega)}{\omega^2} \tag{6}$$

where η_0 is the zero shear viscosity, the ratio of steady state shear stress to shear rate, and J_e^0 is the recoverable shear compliance, the ratio of total recoil strain to shear stress at steady state.[1] Both η_0 and J_e^0 are properties which depend on the slow dynamics. Their product is a mean relaxation time which characterizes the slow dynamics:

$$\tau_m = \eta_0 J_e^0 = \int_{0}^{\infty} tG(t) \, dt \Big/ \int_{0}^{\infty} G(t) \, dt \tag{7}$$

In entangled solutions of linear chains $G(t)$ separates into two rather distinct groups of relaxations, the transition region at short times and the terminal region at long times (Figure 3). The separation begins to appear at a certain chain length and then widens rapidly with increasing chain length. The modulus at intermediate times, the plateau modulus G_N^0, is independent of chain length but depends on polymer species and concentration. The time scale of the terminal region is set by τ_m [Eq. (7)]. In the

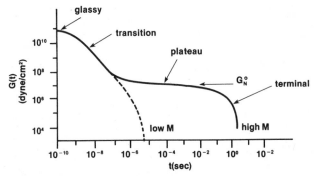

FIGURE 3. Representation of the master curve for shear stress relaxation modulus at high polymer concentrations.

frequency domain (Figure 4) the low-frequency peak in $G'(\omega)$, G_m'' at frequency ω_m, marks the location of the terminal region. In the plateau region $G'(\omega)$ is nearly constant, while $G''(\omega)$ passes through a shallow minimum. Both rise again when the transition region is encountered. The plateau modulus serves, in effect, as the initial modulus for the terminal region. Its value can be estimated either as the nearly constant value of $G'(\omega)$ at intermediate frequencies, or from the resolved area under the terminal loss peak using Eq. (4).[1]

With few exceptions, a change in temperature merely shifts the time scale in the terminal region. The form of the response curve remains the same, and the modulus scale changes little, if at all. This property of time–

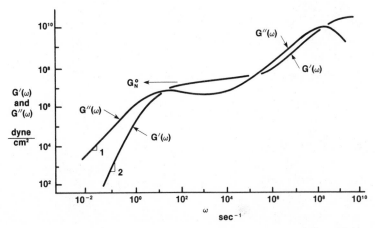

FIGURE 4. Representation of the master curve for dynamic moduli at high polymer concentrations.

temperature superposition is used extensively to reduce data obtained at different temperatures to a master curve at some chosen reference temperature T_0.[1] Thus, for example,

$$G(t) = b_T G_0(t/a_T) \tag{8}$$

where $G(t)$ and $G_0(t)$ are relaxation moduli at temperatures T and T_0, the temperature dependence is contained entirely in the time and modulus shift factors a_T and b_T ($a_T = b_T = 1$ at T_0), and $b_T \sim 1$ within the accuracy of most measurements. When Eq. (8) applies, the behavior is said to be thermorheologically simple, and the temperature dependence of all linear viscoelastic properties is set by a_T and b_T through Eq. (2).[9] Since $G_N{}^0$, $G_m{}''$, and $J_e{}^0$ depend on b_T alone, they are quite insensitive to temperature. Properties such as η_0, ω_m, and τ_m depend strongly on temperature through a_T. Temperature coefficients are governed primarily by the local composition. They vary widely with polymer and solvent species and become independent of chain length for long chains.[1] As a result, the relationships between molecular architecture and properties in the terminal region have forms which are independent of temperature. (Some exceptional behavior in branched polymers of certain species is discussed in Section 6).

4. RELATIONSHIPS WITH MOLECULAR STRUCTURE

4.1. The Plateau Modulus

In undiluted polymers the plateau modulus $G_N{}^0$ is essentially a characteristic constant of the species, being independent of chain length and insensitive to temperature. Well-established values for several species are given in Table 1. With the equation for modulus from the theory of rubber elasticity, these numbers can be recast in terms of the entanglement molecular weight M_e, an apparent molecular weight of the temporary network strands:

$$M_e = \varrho \phi RT/G_N{}^0 \tag{9}$$

where ϱ is the density of undiluted polymer, ϕ is the volume fraction of polymer, and R is the universal gas constant. The variation with species is reduced somewhat when the data are expressed in terms of an entanglement strand length L_e, but even here the values range from about 15 to 60 nm.

TABLE 1. Plateau Modulus for Selected Polymers

Polymer species	T (°C)	$G_N{}^0$ (MPa)	M_e
Polyethylene	190	2.2	1,300
Hydrogenated 1,4-polyisoprene	50	1.2_5	1,800
1,4-Polybutadiene	25	1.1_5	1,900
1,2-Polybutadiene	25	0.62	3,600
1,4-Polyisoprene	25	0.35	6,300
Polyisobutylene	25	0.25	8,900
Poly(dimethyl siloxane)	25	0.24	10,000
Polystyrene	160	0.18	19,500

The variation of $G_N{}^0$ with concentration has been determined in several species. A remarkably general power law dependence on the volume fraction of polymer is found (Figure 5):

$$G_N{}^0 \propto \phi^d \tag{10}$$

The variation appears to be slightly stronger than quadratic ($2.1 < d < 2.3$,

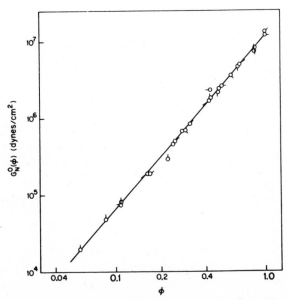

FIGURE 5. Plateau modulus vs. concentration for solutions of 1,4-polybutadiene. The various symbols represent different diluents, including low-molecular-weight polybutadiene.[10] The power law exponent (slope of the line) is 2.26 in this case.

depending on species), insensitive to temperature, and independent of the choice of diluent.[10,11] (However, information on $G_N{}^0$ in θ solvents is still very limited.) Equation (10) appears to apply even as low as $\phi = 0.1$,[10] but resolution of the terminal and transition dispersions is difficult in that range unless the chains are extremely long. Based on the idea that $G_N{}^0$ is proportional to the density of binary contacts one would expect $d = 2$. The slightly larger power is still unexplained. A prediction that $G_N{}^0 \propto \phi^{2.25}$ for semidilute solutions[2] is close to the observation, but the theoretical basis for this seems questionable at high concentrations.

Evaluation of plateau modulus in branched polymers is difficult because long branches broaden the terminal dispersion (see Section 6). However, the existing data suggest that $G_N{}^0$ is changed little, if at all, by long branches.[12] Values of $G_N{}^0$ have been estimated in several species from the topological contribution to the equilibrium modulus of cross-linked networks.[13] These observations all support the idea that $G_N{}^0$ is a fundamental property of the species which transcends the various architectures, reflects the uncrossability constraint, and varies universally with polymer concentration.

Even the observed variation of $G_N{}^0$ with polymer species can be reconciled to some extent.[11] Consider the mechanical behavior of a hypothetical liquid filled with uncrossable molecular "strings" which are very thin, extremely long, and locally flexible. The plateau modulus of this liquid should reflect only the uncrossability interaction between strings, and, like the cross-link density in a rubber network, $G_N{}^0/kT$ should be a measure of the number of interactions per unit volume. This interaction density certainly must depend on the total length of string per unit volume. The contour length concentration is νL_0, ν being the number of strings per unit volume and L_0 the length of each string. If these were the only variables, the relationship between $G_N{}^0/kT$ and νL_0 would be dictated by dimensional considerations alone. The resulting form, however,

$$\frac{G_N{}^0}{kT(\nu L_0)^{3/2}} = \text{pure number} \tag{11}$$

gives the wrong concentration dependence [$G_N{}^0 \propto \phi^{3/2}$ according to Eq. (11), instead of the experimental exponent, $2.1 < d < 2.3$].

There must be at least one other variable in the problem. A natural choice is the Kuhn step length l, characterizing the local flexibility of the chain and thus the distribution of conformations.[14] Both interaction density and contour length concentration can now be made nondimensional with l, and their relationship can be chosen to match the observed power law

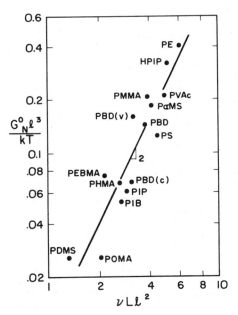

FIGURE 6. Dimensionless plateau modulus vs. contour length concentration for various polymer species in the undiluted state. See Reference 11 for individual species designations.

dependence on concentration. The result,

$$\frac{G_N{}^0 l^3}{kT} = K(\nu L_0 l^2)^d \tag{12}$$

is compared with data for several undiluted polymer species in Figure 6. There is considerable scatter, but the data are reasonably consistent with a quadratic or slightly stronger dependence. For $d = 2$, the average value $K = 0.011$ is obtained. The species dependence and concentration dependence of $G_N{}^0$ can thus be interpreted as separate manifestations of a more general law relating entanglement interactions and uncrossability constraints.

4.2. The Viscosity and Recoverable Compliance

The Rouse model (Chapter 1) provides a useful framework for organizing data on η_0 and $J_e{}^0$. The diffusion coefficient for noninteracting Rouse chains is given by

$$D_r = \frac{kT}{\zeta_0 n} \tag{13}$$

where ζ_0 is the monomeric friction coefficient, and n is the number of monomers in the chain. The molecular friction coefficient is $\zeta_0 n$, so $D_r \propto M^{-1}$. The stress relaxation modulus for a solution of ν noninteracting linear Rouse chains per unit volume is given by

$$G_r(t) = \nu k T \sum_{p=1}^{p_m} \exp\left(-\frac{p^2 t}{\tau_r}\right) \tag{14}$$

in which

$$\tau_r = \frac{\zeta_0 n \langle R^2 \rangle}{6\pi^2 k T} \tag{15}$$

is the Rouse relaxation time, and $\langle R^2 \rangle$ is the mean square end-to-end distance of the chains ($\langle R^2 \rangle = L_0 l$). Both n and $\langle R^2 \rangle$ are proportional to chain length, so $\tau_r \propto M^2$.

Each value of p specifies one of the normal modes of the chain. Qualitatively, τ_r/p^2 is the relaxation time for the equilibration of conformation over a chain distance L_0/p. The upper limit, p_m, designates a species-dependent cutoff at short times. Prior to $t \sim \tau_r/p_m^2$, corresponding to equilibration over distances smaller than L_0/p_m, the response is sensitive to the polymer species. Although Eqs. (14) and (15) are exact for Rouse chains only in the limit $p_m \to \infty$, they are sufficiently accurate for discussing properties which depend primarily on the slow dynamics even for quite modest p_m. The Rouse chain is a one-parameter model for the slow dynamics. Only the molecular friction coefficient $n\zeta_0$ must be established by separate dynamic experiments.

With Eqs. (5) and (6), and for $p_m \gg 1$, Eqs. (14), (15) give the following expressions for viscosity and recoverable compliance:

$$(\eta_0)_r = \frac{\nu \zeta_0 n S^2}{6} \propto M \tag{16}$$

$$(J_e^0)_r = \frac{2}{5} \frac{1}{kT} = \frac{2}{5} \frac{M}{cRT} \tag{17}$$

where $S^2 = \langle R^2 \rangle/6$ is the mean-square radius of gyration of the chains. The experimental observations on undiluted linear chains are as shown in Figures 7 and 8. Below M_c the viscosity behavior agrees fairly well with the Rouse model both in magnitude (using values of ζ_0 estimated from either the transition region response or the diffusion coefficients of small molecules in the polymer[1]) and in chain length dependence. Below M_c' the recoverable compliance is in similarly good agreement. In this case no estimation of ζ_0 is required to make the comparison. The characteristic

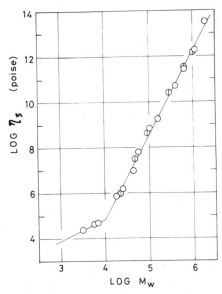

FIGURE 7. Viscosity vs. molecular weight for undiluted polyisoprene. The slope of the line is 1.0 and 3.7, respectively. [See H. Odani, N. Nemoto, and M. Kurata, *Bull. Inst. Chem. Res.* **50**, 117 (1972).]

molecular weights M_c and M_c' mark the boundaries between short-chain behavior [rough consistency with Eqs. (16) and (17)] and long-chain behavior ($\eta_0 \propto M^{3.4}$ and J_e^0 independent of M).

Values of M_e, M_c, and M_c' for several species are given in Table 2. They are insensitive of temperature, and they increase with dilution by what appears to be the same simple law, i.e., they vary as ϕ^{1-d}, d being the dilution exponent for G_N^0 [Eq. (10)]. The relative values, roughly $M_e : M_c : M_c' :: 1 : 2 : 7$, seem to be independent of species and dilution. (There may be some exceptions; experimental uncertainties about the concentration dependence of M_c preclude a definite statement at the present time.)

Combining these observations, one can express the data on η_0 and J_e^0 for solutions of linear, nearly monodisperse chains in the entanglement region as follows:

$$\eta_0(\phi, M) = [\eta_0(\phi, M)]_r \left(\frac{MG_N^0}{2\varrho\phi RT} \right)^{2.4} \tag{18}$$

$$J_e^0(\phi, M) = 2.8/G_N^0 \tag{19}$$

where $M_c/M_e = 2$ and $M_c'/M_e = 7$ have been used, G_N^0 is evaluated at

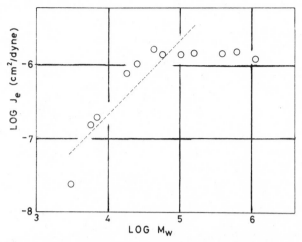

FIGURE 8. Recoverable compliance vs. molecular weight for undiluted polyisoprene (see caption of Figure 7). The dashed line is the Rouse model prediction.

concentration ϕ, and precise agreement with Eqs. (16) and (17) at M_c and M_c' is assumed. The monomeric friction factor is contained in $(\eta_0)_r$. It depends on solvent and polymer species, concentration and temperature but is supposed to be independent of chain length for long chains. Recoverable compliance is extremely sensitive to molecular weight distribution.[1] Even the most carefully prepared samples are not monodisperse, of course, and the numerical coefficient for truly monodisperse chains is probably slightly smaller than that given in Eq. (19).

These universal laws for η_0 and J_e^0 suggest the possibility that the form of $G(t)$ in the terminal region is also universal. That appears to be the

TABLE 2. Characteristic Molecular Weights for Selected Polymers

Polymer species	M_e	M_c	M_c'
Polystyrene	19,500	35,000	130,000
Poly(α-methyl styrene)	13,500	28,000	104,000
1,4-Polybutadiene	1,900	5,000	13,800
Polyvinyl acetate	12,000	24,500	86,000
Poly(dimethyl siloxane)	10,000	24,400	61,000
Polyethylene	1,300	3,800	12,000
1,4-Polyisoprene	6,300	10,000	35,000
Polyisobutylene	8,900	15,200	—

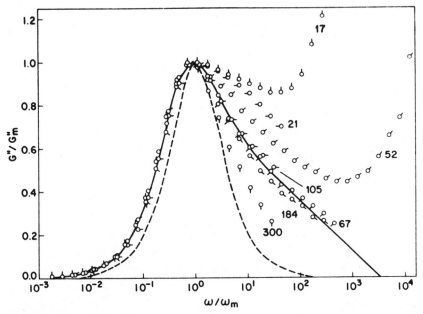

FIGURE 9. Loss modulus vs. frequency in reduced form for several species and chain lengths (see Reference 10 for individual designations). The numbers indicate values of M/M_e. The dashed line is the curve for a single relaxation time: $G''(\omega)/G_m'' = 2(\omega/\omega_m)/[1 + (\omega/\omega_m)^2]$.

case, judged by data on $G^*(\omega)$ at high entanglement densities.[10] Thus, reduced plots of G''/G_m'' vs. ω/ω_m (Figure 9) for various species and concentrations appear to approach a common curve when M/M_c becomes large enough ($M/M_e \sim 100$) to resolve fully the terminal and transition regions.

5. REPTATION AND THE DOI-EDWARDS THEORY

In 1978 Doi and Edwards proposed an attractively simple and testable theory for the slow dynamics of entangled, linear chain liquids.[15–18] They began with the assumption that the chains move from one large-scale conformation to another primarily by reptation, i.e., by diffusing in snakelike fashion along their own contours. de Gennes had introduced this idea earlier when considering the diffusion of unattached chains in a permanent network.[19] Uncrossability constraints from the network tend to suppress lateral motions beyond a certain distance, but motions of the chain as a whole along its own trajectory through the network are unimpeded. In applying this to liquids Doi and Edwards assume that all chains move independently, the role of network now played by neighboring chains, to provide a "tube" for each chain of sufficiently long lifetime.

Before proceeding to the quantitative aspects it is useful to examine some consequences of a purely reptational motion on the chain length dependence of some simple dynamic properties.[19] In any diffusion process the diffusion coefficient, distance and time are related by

$$D = K'X^2/t \tag{20}$$

where K' is a dimensionless constant of order unity. Consider first the lifetime of a large-scale conformation. The chain must reptate a distance of the order of its own length to abandon its current conformation, so $X^2 \propto M^2$. The diffusion coefficient for reptation is inversely proportional to the molecular friction coefficient [see Eq. (13)], so $D \propto M^{-1}$. Thus, from Eq. (20), the conformational lifetime should vary as M^3. Now consider the macroscopic diffusion coefficient for a collection of reptating chains. Each time a chain escapes to a new conformation ($t \propto M^3$), its center of gravity moves a distance of the order of the radius of gyration S. For random coils $S^2 \propto M$, so in this case $X^2 \propto M$. Thus, from Eq. (20), the macroscopic diffusion coefficient should vary as M^{-2}. Note that the diffusion coefficients governing reptation and center of gravity displacement differ. The displacement of the center of gravity produced by a reptational movement is dictated by the current end-to-end vector of the chain.

In the Doi–Edwards formulation the local physical structure of the solution is defined by a mesh size d which depends only on polymer species and concentration. The chains are random coils ($\langle R^2 \rangle = lL_0$), and only the highly entangled regime with local freedom of rearrangement, corresponding to $\langle R^2 \rangle \gg d^2 \gg l^2$, is considered. Each chain is regarded as a connected sequence of random coil segments which is momentarily trapped in an open-ended tube of constraints (Figure 10). It is assumed that the constraint lifetime t_c is effectively infinite, as would be the case in a permanent network.

Local conformations are explored rapidly, and the average of these explorations specifies a trajectory through the mesh, the primitive path of the chain, which defines the current large-scale conformation. The primitive path length L is proportional to chain length L_0 but not as long because the lateral excursions have been averaged out; both are much larger than the mesh size: $L_0 > L \gg d$. Fluctuations are suppressed: L is taken to be

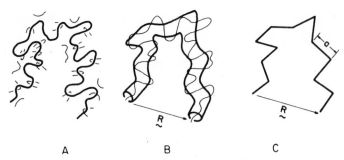

A   B   C

FIGURE 10. Development of the tube model: (A) A chain surrounded everywhere by other uncrossable chains; (B) representation of the uncrossability restriction by a tube; (C) representation of the tube as a random path of step length a with the same end-to-end vector as the chain.

the same for all chains of the same size, and the length of chain per unit length of path, the chain density L_0/L, is constant along the path and has some equilibrium value that depends only on the mesh size. The paths are modeled as random walks with N independently directed steps of step size a, where $N \gg 1$, and a depends only on the mesh size. Thus,

$$Na = L \tag{21}$$

and since the ends of chain and path coincide,

$$Na^2 = \langle R^2 \rangle \tag{22}$$

The chains are not permanently trapped because they can diffuse along their tubes without violating uncrossability. The end of a chain can choose any direction through the mesh as it emerges from its current tube. Primitive path steps are created at the emerging end and abandoned at the other. If the constraint lifetimes are long enough, the orientation of a path step will remain constant from the moment it is created by an emerging chain end until it is abandoned by being visited by either end of the same chain.

The dynamics of the chain in its tube are assumed to be governed by the Rouse model. Thus the diffusion coefficient along the tube is D_r [Eq. (13)], and the time to equilibrate any nonuniformity in the distribution of chain density along the path, t_e, is of the order of τ_r [Eq. (15)]. Solution of the diffusion equation for chains moving along random walk paths give equations for the macroscopic diffusion coefficient:

$$D = D_r/3N \tag{23}$$

and the fraction of path steps with lifetimes longer than time t:

$$F(t) = \frac{8}{\pi^2} \sum_{\substack{odd \\ m}} \frac{1}{m^2} \exp\left(\frac{-m^2 t}{\tau_d}\right) \tag{24}$$

where τ_d is the disengagement time,

$$\tau_d = L^2/\pi^2 D_r \tag{25}$$

Note that $D \propto M^{-2}$ and $\tau_d \propto M^3$, as anticipated by the arguments following Eq. (20). Since $\tau_e \sim \tau_r \propto M^2$, we can expect the chain density along the path to equilibrate quickly compared with the time to reptate out of the tube when the chains are very long. That property is important in the interpretation of mechanical response.

An instantaneous deformation of the system carries the chains into a nonequilibrium distribution of conformations. The resulting increase in configurational free energy generates a restoring stress. The lengths and directions of the primitive path steps are changed, the overall path lengths are increased, and the local density of chain everywhere along the path is displaced from its equilibrium value. If N is large the stress relaxes to zero through a sequence of stages. In stage I the relaxation is dominated by local chain motions. It proceeds rapidly and independently of the uncrossability constraints up to a time of the order of $t_0 = \zeta_0 n \, d^2/6\pi^2 kT$, the Rouse relaxation time for segments of chain with dimensions comparable to the mesh size. Stage I corresponds to the glassy and transition regions (see Figure 3).

In stage II ($t_0 \lesssim t \lesssim t_e$) the chain segments in each path step act like the strands of a random coil network, the stress relaxing slowly as the density of chain equilibrates along the path, and the overall path lengths return to the equilibrium value. At the end of stage II (the plateau region) the orientations along the paths are still distributed according to the original deformation, but the chains are no longer stretched. During stage III (the terminal region) the stress decays from its plateau value to zero as the chains abandon their deformed paths by reptation.

For small strains the stress relaxation modulus in the terminal region is given by

$$G(t) = G_N{}^0 F(t) \tag{26}$$

where $F(t)$ is given by Eq. (24). The relationship follows directly from the property that reptation replaces the oriented path steps at the end of stage

II by randomly diiected steps without altering the orientations of still-occupied parts of the path. The fraction of the plateau stress still remaining is then simply the fraction of steps still occupied. The number of path steps per unit volume is vN (v = chains/volume), so the shear modulus at the beginning of stage II is $vNkT$ from the theory of rubber elasticity. The shear modulus for small strains at the end of stage II is the plateau modulus:

$$G_N{}^0 = \tfrac{4}{5}vNkT \tag{27}$$

The decrease by $1/5$ during stage II comes from the equilibration of local stretches, leaving vN oriented but unstretched random coil segments per unit volume. The combination of Eqs. (26), (27) with Eqs. (5), (6) provides the expressions for viscosity and recoverable compliance:

$$\eta_0 = \frac{\pi^2}{15} vNkT\,\tau_d \tag{28}$$

$$J_e{}^0 = \frac{3}{2} vNkT \tag{29}$$

The Doi–Edwards chain is a two-parameter model for the slow dynamics. The molecular friction coefficient $\eta\zeta_0$ and one additional parameter, characterizing the solution topology, must be supplied by dynamic experiments. Several topological parameters are used (d, L, a, N), but the mesh size d never appears in the equations, and the primitive path length L, step size a, and number of steps N are related through Eqs. (21), (22). Thus, for example, all properties can be expressed in terms of the primitive path step length, which depends only on the polymer species and concentration. From Eqs. (22) and (27),[20]

$$a^2 = \frac{4}{5} \frac{\langle R^2\rangle}{M} \left(\frac{\varrho\phi RT}{G_N{}^0}\right) \tag{30}$$

with $G_N{}^0$ evaluated at the concentration of interest. Assuming the Rouse model holds for solutions of short chains ($M \lesssim M_c$ at the concentration of interest), we can extrapolate, using Eq. (16), to estimate the molecular friction coefficient:

$$n\zeta_0 = \frac{36\eta_0(\phi, M_c)}{\varrho\phi(\langle R^2\rangle/M)} \left(\frac{M}{M_c}\right) \tag{31}$$

With these results the Doi–Edwards equations for diffusion coefficient, viscosity, and recoverable compliance can be written entirely in terms of independently measured quantities:

$$D(\phi, M) = \frac{G_N{}^0}{135} \left(\frac{\langle R^2 \rangle}{M} \right) \frac{M_e{}^2}{M[\eta_0(\phi, M)]_r} \tag{32}$$

$$\eta_0(\phi, M) = 15[\eta_0(\phi, M)]_r \left(\frac{MG_N{}^0}{2\varrho\phi RT} \right)^2 \tag{33}$$

$$J_e{}^0(\phi, M) = 1.2/G_N{}^0 \tag{34}$$

where $M_e = \varrho\phi RT/G_N{}^0$ and $[\eta_0(\phi, M)]_r = \eta_0(\phi, M_c)(M/M_c)$ have been used.

Data on polymer diffusion coefficients in concentrated solutions are rather limited, and especially so under conditions where all the quantities in Eq. (32) can be estimated with confidence. The equation has been tested with diffusion data for undiluted polyethylene,[21] and the agreement in that case is remarkably good. At 176 °C the data can be expressed as $D = 0.34/M^2$ (cm²/sec), and Eq. (32) predicts $D = 0.26/M^2$.[20] This is well within the limits of errors in the measurements (D, $G_N{}^0$, etc.), and certainly better than would be expected from the simplicity of the model itself.

Extensive data on viscosity and recoverable compliance are available for many systems. Comparing Eq. (33) with Eq. (18), expressing the experimental results, one sees that the predicted exponent for viscosity is too small (η_0 proportional to M^3 instead of $M^{3.4}$). However, the numerical coefficient in Eq. (33) makes the predicted magnitude of viscosity too large in the observable range. The situation is shown schematically in Figure 11; the hypothetical crossing point would occur near $M = 800M_c$, well beyond the range of current data. The recoverable compliance is predicted to be independent of chain length, as observed, but its magnitude is too small [compare Eqs. (18) and (34)], as indicated schematically in Figure 11.

What can one conclude then about the assumption that reptation is the dominant mode of motion in entangled solutions of linear chains? Limited data on diffusion coefficients support the assumption, but results for viscosity and recoverable compliance are less clear and imply, at the very least, the existence of processes in addition to reptation. Competing processes would of course result in a more rapid relaxation (smaller viscosity) and a broadened terminal spectrum (larger recoverable compliance), and that is the direction required to explain the observations. Two possible candidates, finite lifetime of the tube constraints and fluctuations in the primitive path length, have been considered.[5,20,22–24]

The tube constraints are supplied by surrounding chains, and, in entangled liquids, each chain is presumably reptating along its own primitive path. It is natural therefore to identify the lifetime of constraints t_c

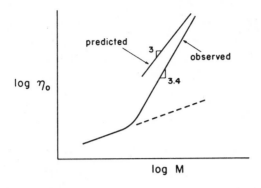

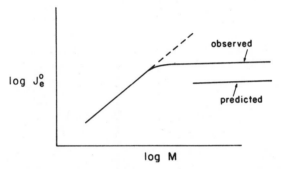

FIGURE 11. Comparison of the Doi–Edwards predictions for viscosity and recoverable compliance with experiment.

in some sense with the mean lifetime of primitive path steps[22]:

$$t_c \approx \int_0^\infty F(t)\, dt = \frac{\pi}{12}\, \tau_d \qquad (35)$$

Release of constraints allows the path locally to shift in position and orientation, thus conferring a Rouse-like random jumping motion on the primitive paths. The need to include constraint release is suggested by the observation that unattached chains relax more slowly in a network environment than in the corresponding entangled liquid.[25] Omission of that contribution may also account for the somewhat unrealistic predictions about polydispersity effects in the Doi–Edwards theory.[5,20]

A simple model has been used to examine the combined effect of reptation and constraint release.[5] It retains the independent chain character of Doi–Edwards; constraint release is treated as a communal property of the solution, acting uniformly on the individual chains. Constraint release turns out to be much less effective than reptation in changing the spatial positions of chains.[5,22] In mixtures the diffusion coefficient is predicted to be insensitive to matrix chain length, as observed experimentally.[26a] Departures from the pure reptation expression [Eq. (23)] should be negligible except for short chains (very small N). The effect of constraint release is predicted to be stronger in stress relaxation. The chain length dependence of η_0 and J_e^0 for large N are unchanged from pure reptation, but the numerical coefficient is smaller for η_0 and larger for J_e^0. As N decreases, constraint release becomes competitive with reptation more rapidly in relaxation than in diffusion. This may account for the observation that the diffusion law for reptation, $D \propto M^{-2}$, appears rather quickly beyond the coil overlap concentration,[26b] while much higher concentrations appear to be necessary to evoke unambiguous entanglement effects in stress relaxation.

The effect of fluctuations in the primitive path length is omitted in the Doi–Edwards theory. Simple reptation deals only with the zeroth mode of motion (translation of the chain as a whole along its path), but the higher modes, the first corresponding to "breathing" of the path, will also result in abandonment of path steps.[23,24] The breathing mode arises from time-dependent fluctuations in chain density along the path. It can be visualized in a crude way as the projection of loops into the surrounding mesh, accompanied by retraction of chain ends toward the center and abandonment of some portion of the path.[5] As the fluctuation subsides, the chain ends then move out again but along random paths (Figure 12).

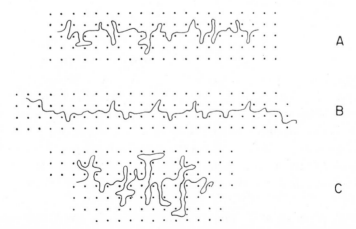

FIGURE 12. Two-dimensional representation of path length fluctuations. (A), (B), and (C) correspond to the same chain at different times and with different instantaneous path lengths.

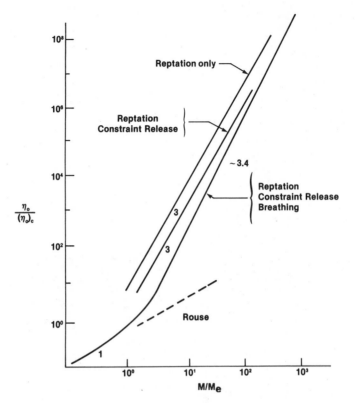

FIGURE 13. Sketch of possible contributions to viscosity from various mechanisms for relaxation.

For linear chains, the contribution of breathing falls off exponentially with chain length,[24] while the rate by reptation goes roughly as $\tau_d^{-1} \propto M^{-3}$. Reptation must then dominate for very long chains, but the contribution from breathing may still be important at intermediate chain lengths. It has been proposed that the experimental law for viscosity, $\eta_0 \propto M^{3.4}$, is not in fact a power law at all but merely the result of a very slow loss of the breathing contribution with increasing chain length, superimposed on reptation.[24] The suggested effects of simple reptation, path length fluctuations and constraint release on viscosity are sketched in Figure 13.

Fluctuations in path length may assume major importance in the relaxation of chains with long branches. Simple reptation would tend to be suppressed by the presence of branch points[23] (see Section 6 following).

6. EFFECTS OF CHAIN BRANCHING

6.1. Experimental Observations

The preceding sections 4 and 5 deal mainly with entanglement effects in solutions of linear chains. These effects are different for branched chains, especially when the branches themselves are long enough to be entangled. The most detailed studies have been made with near-monodisperse star polymers, molecules with three or more arms of equal length joined at a common junction. Each molecule has one branch point, and the branch molecular weight M_b is M/f, f being the number of arms or branch point functionality. There are some data on near-monodisperse comb polymers. Many varieties of randomly branched polymers are available, but even fractionated samples are mixtures of structures, which complicates the interpretation. The discussion here will focus on stars of low functionality ($f = 3, 4, 6$) where crowding near the branch point is not too important. Many of the effects, however, are seen for the other branching architectures as well.

It is useful to discuss the effects of branching in relation to the behavior of linear chains of the same species and at the same concentration.[3,27] In this respect the behavior of branched polymer solutions seems to depend on two structural features, the mean coil size and the length of the branches. If the branches are not too long, the viscosities are rather similar for linear and branched chains with the same radius of gyration. Thus, if $\eta_L(\phi, M)$ is the viscosity for solutions of linear chains, then, approximately,

$$\eta_B(\phi, M) = \eta_L(\phi, gM) \tag{36}$$

where, at the same molecular weight,

$$S_B{}^2 = gS_L{}^2 \tag{37}$$

The size ratio g can be estimated either by dilute solution measurements or by random walk calculations from the known branching architecture. The latter give

$$g = \frac{3f - 2}{f^2} \tag{38}$$

for f-arm stars.

A dependence on coil size alone below the entanglement point is not surprising. At high concentrations the Rouse model works fairly well in that region, and Eq. (16) ($\eta \propto S^2$) is in fact valid for Rouse chains of any architecture. However, Eq. (36) continues to apply even in the entanglement region [where η_L obeys Eq. (18)] as long as the molecular weight of the branches, M_b, is less than M_c.

For longer branches the viscosity begins to exceed the predictions of Eq. (36). The ratio $\eta_B(\phi, M)/\eta_L(\phi, gM)$ starts near unity but increases extremely rapidly with both concentration and branch length. It can easily reach values of 100 or more. The observed behavior of stars in relation to linear chains is sketched in Figure 14. At high molecular weights the viscosity is smaller for stars because the coil size effect is acting alone. At high molecular weights the viscosity becomes larger because the branch length effect is also acting.

Departures from Eq. (36) are expressed in terms of a viscosity enhancement factor,[3,27]

$$\Gamma = \eta_B(\phi, M)/\eta_L(\phi, gM) \qquad (39)$$

The enhancement factor appears to be an exponential function of both arm length and some power of the concentration:

$$\log \Gamma \propto \phi^v M_b \qquad (40)$$

where v is about 1.8. For stars the precise dependence on branch point functionality is unsettled. At constant ϕ and M_b the enhancement clearly increases with f,[27] and this was attributed earlier to an increase in enhancing "efficiency." However, Γ turns out to be about the same for different small functionalities ($f = 3$ and 4) at the same *total* molecular weight $M = fM_b$. Whether that behavior continues at still larger functionalities is not known. Other complications enter the picture at larger f, e.g., chain dimensions are increased, probably because of crowding near the branch point.[28] There

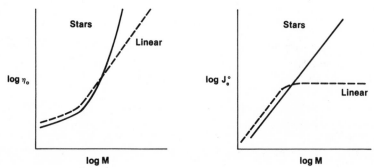

FIGURE 14. Comparison of viscosity and recoverable compliance for liquids of linear chains and star-branched chains.

may also be some reduction of intermolecular entanglement density in the coil interior.[29] At present, the viscosity behavior of stars can be summarized by the equation

$$\eta_B(\phi, M, f) = \eta_L(\phi, gM) \exp[\gamma\phi^v M_b/M_c] \tag{41}$$

where M_c is the characteristic molecular weight for the undiluted species, and γ is a number of order unity which is relatively insensitive to species but varies somewhat with functionality. The rather limited data for combs are also consistent with Eq. (41),[30] g having been obtained by dilute solution measurements.

Like viscosity, the recoverable compliance for branched polymers at low and moderate concentrations is smaller than for linear polymers at the same concentration and molecular weight. For stars, the observed values agree well with the predictions for Rouse stars,[31]

$$J_e^0 = g_2 \frac{2}{5} \frac{M}{\varrho\phi RT} \tag{42}$$

where

$$g_2 = \frac{15f - 14}{(3f - 2)^2} \tag{43}$$

However, in contrast to linear chain behavior, Eq. (42) continues to describe the recoverable compliance for stars even at high concentrations and molecular weights (Figure 14). Thus, above M_c',

$$J_e^0 \propto (G_N^0)^{-1} \propto M^0\phi^{-d} \qquad \text{(linear chains)} \tag{44}$$

while

$$J_e^0 \propto M^1\phi^{-1} \qquad \text{(stars)} \tag{45}$$

with coefficients remarkably close to those predicted by Eq. (42), (43) (see Figure 15). Values of J_e^0 for highly entangled polymers with long branches thus can greatly exceed those for their linear chain counterparts.

The terminal relaxation spectrum for entangled star solutions is broader than the universal form for linear chains.[10] Despite the surprising agreement of J_e^0 with Eq. (42), the terminal region for long arms bears no resemblance to the Rouse form for stars, being much broader and continuing to broaden indefinitely with increasing concentration and arm length.[10]

One curious feature that accompanies viscosity enhancement, but only in some species, is a change in the temperature coefficient of viscosity.

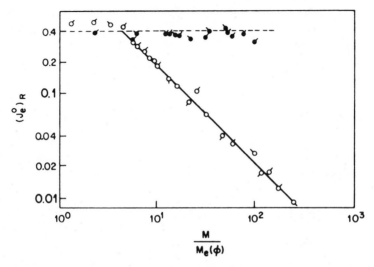

FIGURE 15. Recoverable compliance vs. entanglement density for solutions of linear and star polybutadienes. The compliance values are expressed in reduced form: $(J_e^0)_R = J_e^0 \varrho \phi RT/g_2 M$. For the Rouse model $(J_e^0)_R = 0.4$ (dashed line). The open symbols represent linear polymers; the filled symbols represent branched polymers.

The most prominent example is polyethylene. Temperature effects in this case were studied with model polyethylenes, made by hydrogenation of linear and star polybutadienes.[32] The temperature dependence of viscosity for this polymer is essentially Arrhenius, $\eta_0(T) = A \exp(E_a/RT)$, where E_a is the flow activation energy. For long linear chains, $E_a \sim 7$ kcal for the undiluted polymer, independent of chain length. For three-arm stars, E_a increases in direct proportion to arm length, reaching 15 kcal for $M_b = 30,000$. The difference, $\Delta E = (E_a)_B - (E_a)_L$, decreases smoothly to zero with dilution and generally seems to vary with ϕ and M_b in the same way as the enhancement factor exponent [Eq. (40)]. The shape of the terminal region also changes with temperature when $\Delta E \neq 0$. Polyethylene stars are thermorheologically complex: The short time response, corresponding to small chain distances, has the same temperature coefficient for linear and branched polyethylene, but the temperature coefficient for the longer time response increases progressively in branched polyethylene. In the poly-butadiene precursors, the temperature coefficients of viscosity are virtually independent of chain architecture, and the behavior is thermorheologically simple even when the viscosity enhancement is substantial.[33] Other species show similar effects: ethylene–propylene copolymer stars are thermo-rheologically complex, for example, but polystyrene and polyisoprene stars are thermorheologically simple. The linear chain versions of all these species are thermorheologically simple. The effect appears to be related to the temperature coefficient of chain dimensions[34] and will be considered briefly in the following discussion of relaxation mechanisms in branched polymers.

6.2. Molecular Theory

It seems reasonable to expect that entanglements will reduce the mobility of chains with long branches relative to linear chains. A linear chain, and perhaps even one with sufficiently short branches, can still move freely along its tube of constraints. Interference should develop as the size of the branch becomes comparable to the mesh size, and still longer branches should tend to act as anchors. The tubes themselves are branched, and the uncrossability constraints now resist displacement of the chain as a whole in any direction. Simple reptation is suppressed; diffusion and relaxation must then proceed by other mechanisms such as the fluctuations in primitive path length discussed in Section 5.

The first investigation of fluctuation effects was made by de Gennes.[23] He examined the conformations of a long flexible chain, free at one end but fixed at the other, in a lattice of uncrossable lines. The chain conformation consists of the path through the lattice and nonentangled loops projecting out from the path. The path length varies with the amount of chain in the loops (Figure 16), and, to abandon a path entirely, the chain

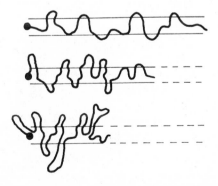

FIGURE 16. Progressive abandonment of an initial path by a chain tethered at one end. The large-scale conformation of the branch relaxes as the result of progressively larger fluctuations in path length.

must form itself into a completely nonentangled loop. de Gennes showed that the probability of finding this conformation decreases exponentially with chain length, implying a conformation lifetime of the form

$$\tau = \tau' \exp(\beta N) \tag{46}$$

where N is the average number of steps in the primitive path, β is a constant of order unity, and the prefactor τ' depends at most on some relatively small power of N.

To test the effect of this on viscoelastic properties, consider the simple example of f-arm star chains ($N_b = N/f$) in which the lifetimes of the path steps vary exponentially with distance from the chain end[†]:

$$\tau_j = \tau' \exp(\beta j) \qquad (1 < j < N_b) \tag{47}$$

The fraction of initial steps which are still occupied at a later time t is the fraction of steps with $t < \tau_j$, or approximately

$$F(t) = \frac{1}{N_b} \sum_{j=1}^{N_b} \exp\left[\frac{-t}{\tau_j}\right] \tag{48}$$

and, with Eqs. (5), (6) and (24), (25),

$$\eta_0 = \frac{4}{5} f\nu kT \sum \tau_j \tag{49}$$

$$J_e^0 = \frac{5}{4f\nu kT} \frac{\sum \tau_j^2}{(\sum \tau_j)^2} \tag{50}$$

For large N_b, the example spectrum [Eq. (47)] gives

$$\eta_0 = \frac{4}{5} \frac{\nu f kT \tau'}{e^\beta - 1} \exp(\beta N_b) \tag{51}$$

$$J_e^0 = \frac{5}{4f\nu kT} \frac{e^\beta - 1}{e^\beta + 1} \tag{52}$$

Thus, as observed experimentally, the viscosity grows exponentially with arm length, and recoverable compliance has the Rouse form, $J_e^0 \propto \nu^{-1}$:

$$J_e^0 = \left[\frac{25}{8f} \frac{e^\beta - 1}{e^\beta + 1}\right] \frac{2}{5} \frac{M}{\rho \phi RT} \tag{53}$$

[†] I am grateful to Dr. Robin Ball of Cambridge University for suggesting this illustration of the principle.

Even the coefficient in brackets is similar in magnitude to the Rouse–Ham g_2 (Eq. (43)): $g_2 = 0.63$ for $f = 3$, and (with $\beta = 1$) the bracketed coefficient is 0.48.

The principle illustrated by these calculations is general even though the example spectrum [Eq. (47)] is probably incorrect in detail. The terminal spectrum for disengagement by fluctuations must be very broad compared with that for simple reptation. The viscosity in such cases is dominated by the longest relaxation time, and that will certainly be exponential in arm length. Moreover, in the context of tube models generally, spectral broadness will inevitably lead to a Rouse-like form for the recoverable compliance. Equation (50) for J_e^0 is valid for tube models regardless of the spectral form. For linear chains (simple reptation) the step lifetimes are all of comparable magnitude, so $\sum \tau_j^2 / (\sum \tau_j)^2 \propto N/N^2 = N^{-1}$, and $J_e^0 \propto (\nu N)^{-1} \propto (G_N^0)^{-1}$, the Doi–Edwards form and the observed behavior for entangled linear chains [Eq. (19)]. For long arm stars the step lifetimes increase rapidly with distance from the free end. The summations in such cases are dominated by their largest terms, so $\sum \tau_j^2 / (\sum \tau_j)^2$ is independent of N, and $J_e^0 \propto \nu^{-1} \propto M^1 \phi^{-1}$. A wide spacing of the longest relaxation times is enough to produce the Rouse dependence on ϕ and M; no further resemblance to the Rouse spectrum is required.

Disengagement by fluctuations in path length also provides a possible explanation for thermorheological complexity in the branched polymers of certain species.[34] The transition states for relaxation involve increasingly more compact (highly looped) chain conformations for parts of the path near the branch point, requiring an increasingly larger proportion of *gauche* conformers and correspondingly fewer *trans* conformers. When these conformers have the same energy, the excess free energy of the transition state will be purely entropic. When the *gauche* conformers have higher energy, the chains must pass through states of higher energy as well as lower entropy. The temperature dependence of relaxation in the latter case will vary with distance from the branch point. Thus, the parameter β [Eq. (46)] should depend on both temperature and location along the branch if $\Delta \varepsilon$, the energy difference between *gauche* and *trans* isomers, is greater than zero for the species. Moreover, it should have the form

$$\beta_j(T) = \beta_0 + \frac{E^\ddagger}{RT} j \qquad (54)$$

assuming that the excess of *gauche* conformers in each transition state is proportional to the surplus of chain to be distributed as loops.

In polymer species with $\Delta\varepsilon > 0$, the spectrum of stars should narrow with increasing temperature (thermorheologically complex behavior). The viscosity, dominated by the longest relaxation time, should have a temperature dependence which exceeds that for linear chains by an Arrhenius term with activation energy $\Delta E = E^{\ddagger} N_b$. The effect should therefore increase with branch length and decrease with dilution, as observed.[32] In species with $\Delta\varepsilon \sim 0$, the viscosity temperature coefficient for linear and branched chains should be the same and time–temperature superposition should be obeyed.

The *gauche–trans* energy difference also controls the temperature dependence of chain dimensions for the species: $d\ln C_{\infty}/dT = \varkappa$, $\varkappa$ being negative when $\Delta\varepsilon$ is positive.[6] For polyethylene, in which stars exhibit thermorheological complexity, $\varkappa$ is indeed negative and fairly large compared with most other species. For polybutadiene (precursors of the model polyethylenes), in which stars are thermorheologically simple, the value of $\varkappa$ is nearly zero. These examples and others[34] are consistent with the idea that stars relax primarily by fluctuations in path length, and that differences among species are related to conformational characteristics such as $d\ln C_{\infty}/dT$. However, some part of the effects may be caused by other factors (see Reference 34). In addition, it is clear that constraint release must also be considered in the relaxation of branched chains,[25] so the quantitative aspects are probably more complicated than suggested by Eqs. (46) and (54).

7. NONLINEAR VISCOELASTICITY

The discussion thus far has only dealt with entanglement effects on the slow dynamics of solutions which are essentially at equilibrium. For sufficiently small or sufficiently slow deformations the response is linear and governed by $G(t)$ alone [Eq. (2), for example]. Combinations of large strains and strain rates displace the physical structure of the solution from equilibrium, and the response is no longer linear. Chain conformation and entanglement determine the stress at each instant and these now vary, depending on the prior sequence of strains (the deformation history). The transient network theory of Lodge[35] accounts for many general features of the nonlinear response in entangled solutions, but it requires certain functions as input (the strand production and destruction rates) which must be determined empirically. The BKZ equation[36] is another very useful form but lacking in explicit molecular content. There are numerous continuum formulations and many mathematically acceptable ways to build in departures from linearity.[37] Some fail on empirical grounds alone, but even the remaining candidates suffer from the Lodge and BKZ model limitations. It is not clear how to calculate the functions they require from theories about the chain motions.

Flow instabilities and instrumental problems are more troublesome in nonlinear experiments.[38] Much careful work has been done, but there is still a serious lack of data on nonlinear response for polymers with well-defined molecular structure. The only property which has been thoroughly studied with near-monodisperse systems is the shear-rate dependence of viscosity, i.e., the ratio of stress to shear rate at steady state in simple shear flow: $\eta(\dot{\gamma}) = \sigma(\dot{\gamma})/\dot{\gamma}$. Another nonlinear property, the shear stress relaxation modulus for finite strains, $G(t, \gamma) = \sigma(t, \gamma)/\gamma$, has also received attention recently, and information on this property for well-characterized polymers is accumulating rapidly. The discussion here will focus on these two properties. Molecular effects for other components of the stress tensor at steady state are well documented only at low shear rates. Data suitable for molecular interpretation on stress transients are fragmentary for shear flows and practically nonexistent for extensional flows.

The main features of shear rate dependence of viscosity in highly entangled solutions are now well established.[4] Beyond some range of shear rates, defined by a characteristic shear rate $\dot{\gamma}_0$, the viscosity begins to decrease from its zero shear rate value η_0. Expressed in reduced form, $\eta(\dot{\gamma})/\eta_0$ vs. $\dot{\gamma}/\dot{\gamma}_0$, the behavior appears to be universal for highly entangled polymers with narrow molecular weight distributions (Figure 17). The effects of concentration, molecular weight, temperature, species, and chain architecture (linear or star[39]) are absorbed entirely in the values of η_0 and $\dot{\gamma}_0$. Moreover, $\dot{\gamma}_0$ appears to depend only on the product $\eta_0 J_e^0$, the mean relaxation time τ_m of the slow dynamics [Eq. (6)]:

$$\dot{\gamma}_0 = K'/\eta_0 J_e^0 \qquad \text{or} \qquad \dot{\gamma}_0 \tau_m = K' \qquad (55)$$

where K' is a universal constant whose numerical value depends on the particular method chosen to define $\dot{\gamma}_0$ ($K' \sim 0.5$ when $\dot{\gamma}_0$ is the shear rate at which the viscosity has fallen to $0.8\eta_0$ [4]). Thus

$$\frac{\eta(\dot{\gamma})}{\eta_0} = V(\dot{\gamma}\eta_0 J_e^0) \qquad (56)$$

where $V(\dot{\gamma}\eta_0 J_e^0)$ is a universal function for highly entangled monodisperse solutions. At shear rates well beyond $\dot{\gamma}_0$ this reduced viscosity function

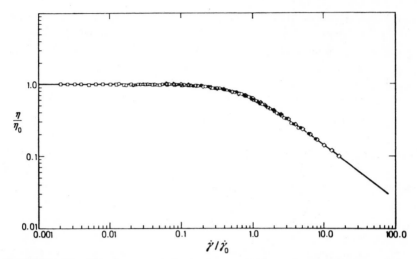

FIGURE 17. Viscosity–shear rate behavior of entangled, nearly monodisperse polymers, expressed in reduced form. Solutions of polystyrene with various chain lengths, concentrations and temperature are represented.[4] Data for other entangled polymer systems, linear and branched, obey the same form. The solid line was calculated from the theory in Reference 40.

goes over to a power law

$$\eta(\dot{\gamma}) \propto \eta_0(\dot{\gamma}\eta_0 J_e^0)^{-p} \tag{57}$$

where the exponent p in the range 0.80–0.85. The behavior at still higher shear rates is uncertain; flow instabilities and viscous heating invariably intervene at some point, although there is some evidence that the viscosity approaches a limiting value.

A surprisingly accurate prediction of the reduced viscosity function has been obtained from a rather simple model for the loss and replacement of chain entanglements at steady state.[40] The entanglements along a chain are lost as the neighboring chains are carried out of its pervaded volume by the flow. New entanglements are formed as new chains arrive, but it is assumed that they are not formed instantly: An incoming chain must remain in the pervaded volume longer than the time for complete rearrangement of its conformation. If it is swept by more quickly, no entanglement occurs, and the resistance to flow is correspondingly diminished. Thus, the steady state entanglement density decreases with increasing shear rate, and the viscosity decreases progressively from its zero shear value η_0. The conformational relaxation time is assumed to vary with shear rate in the same way as the viscosity: $\tau(\dot\gamma) = \tau_0 \eta(\dot\gamma)/\eta_0$, where τ_0 is the relaxation time at rest $(\dot\gamma = 0)$. The resulting viscosity function for monodisperse chains, $\eta(\dot\gamma)/\eta_0$ vs. $\dot\gamma\tau_0$, fits data over a wide range of shear rates, and it goes finally to a power law [Eq. (57)] for $\dot\gamma\tau_0 \gg 1$ with $p = 9/11 = 0.82$, in good agreement with observations. The interpretation of τ_0 as a relaxation time for conformation rearrangement is of course consistent with Eq. (55) but not unique. In general terms it means simply that departures from η_0 begin at shear rates of the order of the reciprocal of the longest relaxation time in the linear viscoelastic spectrum.

As noted earlier, the observed viscosity function is the same for highly entangled solutions of linear and star chains. Their relaxation spectra are entirely different, so it is clear that theories which postulate a direct and universal connection between linear response curves and viscosity-shear rate behavior must be wrong in principle.

The strain-dependent relaxation modulus $G(t, \gamma)$ reduces to $G(t)$, the relaxation modulus of linear viscoelasticity, for small strains. Departures from $G(t)$ develop as γ increases. The ratio $G(t, \gamma)/G(t)$ decreases from unity by a factor that in general depends both on strain and time. In entangled solutions there are two time domains,[41] separated by t_k, a characteristic time for the solution. Prior to t_k the ratio depends on both strain and time. Beyond t_k, the ratio is a function of strain alone: The strain and time dependences are approximately factorable (Figure 18). Thus

$$G(t, \gamma) = h(\gamma)G(t) \qquad (t > t_k) \qquad (58)$$

The time t_k lies in the plateau region for the solution, so t_k is small compared with the longest relaxation times in entangled solutions, and Eq. (58) describes the terminal response for finite strains. The strain-dependent function $h(\gamma)$ appears to be relatively insensitive to polymer species, concentration and chain architecture at moderate entanglement densities, roughly from $M = 5M_e(\phi)$ to $M = 50M_e(\phi)$. Below $5M_e(\phi)$ the value of t_k becomes comparable to the longest relaxation time; the explorations of behavior here are quite limited. Above $\sim 50M_e(\phi)$ the factorability of strain and time dependence becomes less precise, and the shifts with strain become progressively larger and somewhat erratic. The reason for the existence of an upper limit is unknown.

The Doi–Edwards theory provides a natural explanation for some of these observations.[16] In the plateau and terminal regions, the relaxation following large strains is expected to take place in two stages which are well separated in time scale for long chains. The faster process is equilibra-

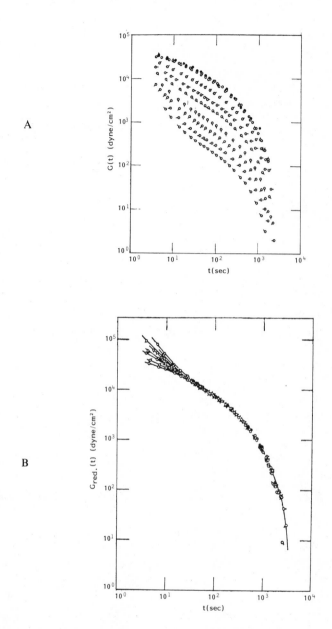

FIGURE 18. (A) Stress relaxation curves for various shear strains for a 20% solution of high-molecular-weight polystyrene with narrow molecular weight distribution.[43] Up to $\gamma = 1.87$ the values of $\sigma(\gamma, t)/\gamma$ coincide (top curve); beyond that (up to $\gamma = 25.4$) they decrease progressively, but maintain the same shape beyond $t_k \sim 10$ s. (B) Data in Figure 18(A) shifted along the modulus axis to achieve superposition with the upper (linear viscoelastic) curve. The shift factor obtained in this way is $h(\gamma)$.

tion of stretches along the path and return of the path length to its value at equilibrium ($t \sim t_e$, the equilibration time). The slower process is replacement of oriented path steps by randomly directed steps by reptation ($t \sim \tau_d$, the disengagement time). The characteristic time t_k would correspond with the equilibration time t_e, and should therefore be of the order of the Rouse relaxation time $\tau_r \propto M^2$. Experimentally, t_k varies approximately as M^2, and its values are indeed of the order of estimations based on Eq. (15).[41]

The strain-dependent function $h(\gamma)$ is accordingly the fraction of the plateau modulus $G_N{}^0$ remaining after equilibration. Part of the reduction comes from the abandonment of path steps as the chain ends retract into the tube and part from the equilibration of stretches in the remaining steps.[16] The former corresponds to a decrease in entanglement density, at least in the sense that the concentration of path steps is smaller than at equilibrium. However, the theory predicts there should be no effect on the time dependence of subsequent disengagement: the path length, and hence the reptation distance, is already at its equilibrium value. Therefore, the time and strain dependences should be factorable beyond t_e, as observed, and the terminal region time dependence should be given by the linear response function $G(t)$, as in Eq. (58).

Doi and Edwards derived an equation for the strain-dependent function assuming nothing more than affine displacements and return to the original path length after equilibration.[16] The form of this function should be universal for entangled linear and star chains. The expression obtained for $h(\gamma)$ is in excellent agreement with experimental data.[41] The theory of course supplies no explanation for the observed departures from Eq. (58) when $M \gtrsim 50 M_e \phi$.

The tube model has been generalized to arbitrary deformation histories,[17] but only by the introduction of additional assumptions. The resulting expression for $\eta(\dot{\gamma})$ does not agree with experiment.[18] The decrease with shear rate is too rapid; indeed, the shear stress at steady state is predicted to pass through a maximum with shear rate, and this is clearly inconsistent with the data. It is possible, however, that the problem is not with the tube model itself, but with the simplifying assumption that equilibration is instantaneous. Measurements on other nonlinear properties suggest that departures from the Doi–Edwards predictions are correlated with the equilibration time of the solution.[42] A generalized theory which includes both equilibration and reptation contributions will be required to test that possibility.

REFERENCES

1. J. D. Ferry, *Viscoelastic Properties of Polymers*, 3rd ed., John Wiley and Sons, New York (1980).
2. P. G. de Gennes, *Scaling Concepts in Polymer Physics*, Cornell University Press, Ithaca, New York (1979).
3. G. C. Berry and T. G. Fox, *Adv. Polym. Sci.* **5**, 261 (1968).
4. W. W. Graessley, *Adv. Polym. Sci.* **16**, 1 (1974).
5. W. W. Graessley, *Adv. Polym. Sci.* **47**, 67 (1982).
6. P. J. Flory, *Statistical Mechanics of Chain Molecules*, Interscience, New York (1969).
7. G. C. Berry, H. Nakayasu, and T. G. Fox, *J. Polym. Sci. Polym. Phys.* **17**, 1825 (1979); W. W. Graessley, *Polymer* **21**, 258 (1980); B. L. Hager and G. C. Berry, *J. Polym. Sci. Polym. Phys.* **20**, 911 (1982).
8. W. H. Stockmayer, in *Molecular Fluids, Les Houches 1973* (B. Balian and G. Weill, eds.), pp. 107–144, Gordon and Breach, London (1976).
9. H. Markovitz, *J. Polym. Sci. Polym. Symp.* **50**, 431 (1975); D. J. Plazek, E. Riande, H. Markovitz, and N. Raghupathi, *J. Polym. Sci. Polym. Phys.* **17**, 2189 (1979).
10. V. R. Raju, E. V. Menezes, G. Marin, W. W. Graessley, and L. J. Fetters, *Macromolecules* **14**, 1668 (1981).
11. W. W. Graessley and S. F. Edwards, *Polymer* **22**, 1329 (1981).
12. W. W. Graessley and J. Roovers, *Macromolecules* **12**, 959 (1979).
13. S. Granick and J. D. Ferry, *Macromolecules* **16**, 39 (1983), and references therein.
14. P. J. Flory, *Principles of Polymer Chemistry*, p. 412, Cornell University Press, Ithaca, New York (1953).
15. M. Doi and S. F. Edwards, *J. Chem. Soc. Faraday Soc. Trans.* 2 **74**, 1789 (1978).
16. M. Doi and S. F. Edwards, *J. Chem. Soc. Faraday Soc. Trans.* 2 **74**, 1802 (1978).
17. M. Doi and S. F. Edwards, *J. Chem. Soc. Faraday Soc. Trans.* 2 **74**, 1818 (1978).
18. M. Doi and S. F. Edwards, *J. Chem. Soc. Faraday Soc. Trans.* 2 **75**, 38 (1979).
19. P. G. de Gennes, *J. Chem. Phys.* **55**, 572 (1971).
20. W. W. Graessley, *J. Polym. Sci. Polym. Phys.* **18**, 27 (1980).
21. J. Klein and B. J. Briscoe, *Proc. R. Soc. (London)* **365**, 53 (1979).
22. J. Klein, *Macromolecules* **11**, 852 (1978); M. Daoud and P. G. deGennes, *J. Polym. Sci. Polym. Phys.* **17**, 771 (1981).
23. P. G. de Gennes, *J. Phys.* **36**, 1199 (1975).
24. M. Doi and N. Y. Kuzuu, *J. Polym. Sci. Lett.* **18**, 775 (1980); M. Doi, *J. Polym. Sci. Lett.* **19**, 265 (1981).
25. H. C. Kan, J. D. Ferry, and L. J. Fetters, *Macromolecules* **13**, 1571 (1980).
26a. J. Klein, *Philos. Mag.* **A43**, 771 (1981).
26b. L. Legér, H. Hervet, and F. Rondelez, *Macromolecules* **14**, 1732 (1981).
27. W. W. Graessley, *Acc. Chem. Res.* **10**, 332 (1977).
28. R. Roovers, N. Hadjichristidis, and L. J. Fetters, *Macromolecules* **16**, 214 (1983); M. Daoud and J. P. Cotton, *J. Phys.* **43**, 531 (1982).
29. J. Carella, Ph. D. thesis, Northwestern University, Evanston (1983).
30. J. Roovers and W. W. Graessley, *Macromolecules* **14**, 766 (1981).
31. J. S. Ham, *J. Chem. Phys.* **26**, 625 (1957).
32. W. W. Graessley and V. R. Raju, *J. Polym. Sci. Polym. Symp.* **71**, 77 (1984).
33. W. E. Rochefort, G. G. Smith, H. Rachapudy, V. R. Raju, and W. W. Graessley, *J. Polym. Sci. Polym. Phys.* **17**, 1197 (1979).

34. W. W. Graessley, *Macromolecules* **15**, 1164 (1982).
35. A. S. Lodge, *Body Tensor Fields in Continuum Mechanics*, pp. 142–143, Academic Press, New York (1974); see also M. H. Wagner, *Rheologica Acta* **18**, 33 (1979).
36. B. Bernstein, E. A. Kearsley, and L. J. Zapas, *Trans. Soc. Rheology* **7**, 391 (1963); see also B. Bernstein and R. L. Fosdick, *Rheologica Acta* **9**, 186 (1970); L. J. Zapas and J. C. Phillips, *J. Res. Natl. Bur. Stand. A* **75**, 33 (1971).
37. R. B. Bird, R. C. Armstrong, and O. Hassager, *Dynamics of Polymeric Liquids*, Vol. 1, John Wiley and Sons, New York (1977).
38. J. Meissner, *Rheologica Acta* **10**, 230 (1971); J. Meissner, *J. Appl. Polym. Sci.* **16**, 2877 (1972); R. C. Crawley and W. W. Graessley, *Trans. Soc. Rheol.* **21**, 19 (1977).
39. W. W. Graessley, T. Masuda, J. E. L. Roovers, and N. Hadjichristidis, *Macromolecules* **9**, 127 (1976).
40. W. W. Graessley, *J. Chem. Phys.* **47**, 1942 (1967).
41. K. Osaki and M. Kurata, *Macromolecules* **13**, 671 (1980); C. M. Vrentas and W. W. Graessley, *J. Rheol.* **26**, 359 (1982).
42. E. V. Menezes and W. W. Graessley, *J. Polym. Sci. Polym. Phys.* **20**, 1817 (1982).
43. Y. Einaga, K. Osaki, M. Kurata, S. Kimura, and M. Tamura, *Polym. J. (Japan)* **2**, 550 (1971).

4

Neutron Scattering from Macromolecules in Solution

JULIA S. HIGGINS and ANN MACONNACHIE

1. INTRODUCTION

1.1. General Background

Neutron scattering spectroscopy[1-4] differs from scattering of electromagnetic radiation (light or x-rays) in two major ways. The relatively larger neutron mass, which associates a sizeable momentum transfer with a scattering event, totally changes the relationship between energy and wave vector. This property, which means, for example, that neutrons have very much smaller energies than x-rays of the corresponding wavelength, allows exploration of a unique region of the spatial and time domains. It is, however, doubtful whether this property would have led, alone, to the widespread use by polymer scientists of neutron spectrometers, confined as these are to a few reactor centers scattered worldwide, if it were not for the second property—the neutron–nuclear interaction. Since the neutron is uncharged it interacts with the nucleus via nuclear forces. It carries a magnetic moment which can also interact with the nucleus and with the

JULIA S. HIGGINS and ANN MACONNACHIE ● Department of Chemical Engineering and Chemical Technology, Imperial College, London SW7 2BY, United Kingdom.

unpaired electrons in a molecule. This magnetic scattering is relatively weak and does not concern us when dealing with polymer solutions. The nuclear interaction is strong, but very short range, and thermal neutrons have wavelengths very much larger than nuclear dimensions. For an isolated stationary nucleus, scattering is, therefore, spherically symmetrical and energy independent and can be characterized by a single parameter, the scattering length b. Values of b vary randomly from nucleus to nucleus, from isotope to isotope, and even with the spin state of the scattering nucleus. In particular, the values for ^1H and ^2D are of opposite signs. H–D substitution thus makes relatively simple the labelling of molecules or parts of molecules. It is this property which has led to the importance of the neutron as a probe for investigating conformation and dynamics of polymer molecules.[2a,3,4]

   The distance scale explored experimentally is determined by Q^{-1}, where $\hbar Q$, the momentum transfer, is given for elastic and quasielastic scattering by $Q = 4\pi/\lambda \sin \theta/2$. λ is the neutron wavelength and θ the angle of scatter. Most neutron sources until recently have been thermal reactors which produce a Maxwellian distribution of velocities and hence a wavelength spread centered around 1–2 Å. A number of reactors have cold sources installed. These are containers of liquid hydrogen or deuterium in the exit beam path close to the core of the reactor. Neutrons are equilibrated at the low temperature of the liquid by a number of scattering events and emerge with a distribution shifted to longer wavelengths. Wavelengths of 10 Å and longer are commonly used from such sources. This is particularly important for macromolecular samples as it allows small angle scattering experiments ($\theta \ll 2°$) to explore dimensions up to several hundred angströms. The whole range, from molecular dimensions to quite local structure and conformation, is thus accessible, while H–D exchange allows parts of the system of interest to be labeled. It should be mentioned here that although the scope of this chapter has been limited to results from synthetic polymers, the techniques described have been applied very successfully[5–7] to many biological systems, where the ubiquitous water component has made labeling simple and very valuable.

   The time or frequency scale of molecular motion in the scattering system is observed via energy transfer, to or from the neutron, during scattering. In normal solvents, internal motions of macromolecules range from frequencies above 10^{12} Hz for rotation and vibration of side groups, through local segmental motion at around 10^9 Hz, to reorganization of very long sections of chain backbone at around 10^6 Hz and below. While molecular spectroscopy is a widely used application of neutron scattering, the study of side chain motion of polymers in solution has not, in general, required resort to neutron scattering. The characteristic long-range motion of the polymer backbone—the Rouse motion—has been of considerable interest. However, the energy of long wavelength neutrons is about 1 meV, while the 10^7–10^9 Hz range of even the highest-frequency Rouse modes corresponds to 0.01 to 1 μeV. Thus extremely high-resolution quasielastic scattering is demanded for the study of these types of polymer dynamics using neutrons.

1.2. Coherent and Incoherent Scattering

The scattering cross section, σ, is defined as the ratio of the number of scattering events per second to the incident neutron flux (defined as neutrons per square centimeter per second). It can be seen that σ has dimensions of area and for an isolated stationary nucleus, $\sigma = 4\pi b^2$, where b is the scattering amplitude already mentioned.

In a sample containing nuclei with spin (or for that matter simply an isotopically impure sample) b will vary from site to site. The differential cross section per atom with respect to angle Ω contains information about spatial correlation within the sample as follows:

$$\frac{d\sigma}{d\Omega} = \frac{1}{N} \left| \sum_R b_R \exp i(\mathbf{Q} \cdot \mathbf{R}) \right|^2 \tag{1}$$

where $\mathbf{R}$ is a position vector and N the number of atoms. This expression can be manipulated[1,2] to give

$$\frac{d\sigma}{d\Omega} = \{\langle b^2 \rangle - \langle b \rangle^2\} + \langle b \rangle^2 \left| \sum_R \exp(\mathbf{Q} \cdot \mathbf{R}) \right|^2 \tag{2}$$

where $\langle \ \rangle$ implies averaging over all sites. Only the last term contains spatial information. It is the coherent scattering, and the coherent scattering cross-section ($\sigma_{coh} = 4\pi b^2$) depends on the mean square scattering amplitude. The first term contains no spatial information and is rather uninteresting in a static experiment. It can also be written as $\langle (b - \langle b \rangle)^2 \rangle$, the mean square deviation of the scattering lengths from their average value. This is the incoherent scattering term which in a static experiment forms a flat background to the coherent scattering. For nuclides with no spin σ_{inc} $(= 4\pi \langle (b - \langle b \rangle)^2 \rangle)$ is zero (e.g., ^{12}C, ^{16}O). For hydrogen it is very large with a value of 80×10^{-24} cm^2 compared with values of 1 or 2×10^{-24} cm^2 for other nuclei.

TABLE 1. Values of the Scattering Lengths and Cross-sections of Common Atoms

	$b \times 10^{12}$ cm	$\sigma_{coh} \times 10^{24}$ cm^2	$\sigma_{inc} \times 10^{24}$ cm^2	$\sigma_{abs} \times 10^{24}$ cm^2 (at 1.08 Å)
^{1}H	-0.374	1.76	80	0.19
^{2}D	0.667	5.59	2	0.0005
^{12}C	0.665	5.56	0	0.003
^{14}N	0.94	11.1	0.3	1.1
^{16}O	0.58	4.23	0	0.0001
^{19}F	0.56	3.94	0.06	0.006
^{ave}Si	0.42	2.22	0	0.06
^{32}S	0.28	0.99	0	0.28
$^{ave\ 35.5}$Cl	0.96	11.58	3.5	19.5

Table 1 lists values of b, σ_{inc}, and σ_{coh} together with the absorption cross-section σ_{abs} for nuclei commonly found in synthetic polymers and their solvents.

1.3. The Scattering Laws

In an experiment changes in energy of the scattered nuclei are observed as a function of scattering angle and expressed in terms of the double differential scattering cross-section of the sample per atom, with respect to energy E and angle Ω, $d^2\sigma/d\Omega\,dE$. In general, there will be both coherent and incoherent terms in the scattering and although we express them separately they may be very difficult to separate experimentally. The incoherent cross-section will contain no spatial correlations but may carry information about molecular motion. The coherent cross-section contains information about spatially correlated molecular motion. In the classical limit

$$\frac{d^2\sigma_{coh}}{d\Omega\,dE} = \frac{k'}{k}\frac{\langle b \rangle^2}{2\pi\hbar}\int\int d\mathbf{R}\,dt\,\exp[i(\mathbf{Q}\cdot\mathbf{R} - \omega t)]G(\mathbf{R}, t) \qquad (3)$$

$$\frac{d^2\sigma_{inc}}{d\Omega\,dE} = \frac{k'}{k}\frac{(\langle b^2 \rangle - \langle b \rangle^2)}{2\pi\hbar}\int\int d\mathbf{R}\,dt\,\exp[i(\mathbf{Q}\cdot\mathbf{R} - \omega t)]G_s(\mathbf{R}, t) \qquad (4)$$

$\mathbf{k}$, $\mathbf{k}'$ are the initial and final wave vectors of the neutron so that

$$\mathbf{Q} = \mathbf{k}' - \mathbf{k} \quad \text{and} \quad \delta E = \frac{\hbar^2}{2m}(k'^2 - k^2) = \hbar\omega \qquad (5)$$

It is usual to express the integrals in Eqs. (3) and (4) as scattering laws which contain all the information about the scattering sample

$$S(\mathbf{Q}, \omega) = \tfrac{1}{2}\pi \int\int \exp[i(\mathbf{Q} \cdot \mathbf{R} - \omega t)]G(\mathbf{R}, t)\, d\mathbf{R}\, dt \qquad (6)$$

where $G(\mathbf{R}, t)$ is the space-time correlation function expressing correlations between a scattering center at time t_0 and any center (including itself) at time t:

$$S_s(\mathbf{Q}, \omega) = \tfrac{1}{2}\pi \int\int \exp[i(\mathbf{Q} \cdot \mathbf{R} - \omega t)]G_s(\mathbf{R}, t)\, d\mathbf{R}\, dt \qquad (7)$$

where the self-correlation function $G_s(\mathbf{R}, t)$, as its name suggests, only expresses correlations between a center at time t_0 and itself at time t.

Historically neutron scattering experiments have been used to observe energy changes of the neutrons, leading to the cross-section defined above. Molecular motion can equally be observed as a time correlation function, i.e., in the time rather than the frequency domain (for example, photon correlation spectroscopy). In this case, the intermediate, or time correlation functions $S(\mathbf{Q}, t)$ are observed. These are the time Fourier transforms of $S(\mathbf{Q}, \omega)$ and $S_s(\mathbf{Q}, \omega)$ so that

$$S(\mathbf{Q}, t) = \int \exp(i\mathbf{Q} \cdot \mathbf{R})\, d\mathbf{R}\, G(\mathbf{R}, t) \qquad (8)$$

$$S_s(\mathbf{Q}, t) = \int \exp(i\mathbf{Q} \cdot \mathbf{R})\, d\mathbf{R}\, G_s(\mathbf{R}, t) \qquad (9)$$

The recently developed neutron spin-echo technique described in Section 2.2.2 allows observation of $S(\mathbf{Q}, t)$ directly, and this has proved very useful when dealing with scattering from polymers in solution.

2. EXPERIMENTAL TECHNIQUES

As outlined in Section 1.1 the properties of polymeric solutions which have been subjected to extensive neutron scattering investigation are long-range conformation and backbone dynamics. The former requires small-angle scattering, the latter high-resolution quasielastic scattering techniques. These two neutron methods will be described in some detail in this section.

2.1. Small-Angle Neutron Scattering

An elastic scattering experiment consists of scattering neutrons of known wavelength off the sample under investigation and measuring the intensity of the scattered neutrons as a function of scattering angle. No energy analysis is carried out. A small-angle neutron scattering (SANS)

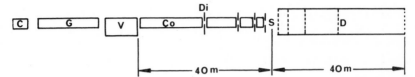

FIGURE 1. Diagram of the small-angle diffractometer D11 at the ILL, Grenoble, illustrating the main features. C is the cold source, G the neutron guide, V the velocity selector, Co one of the four sections of collimation, Di one of the diaphragms, S is the sample position, and D is the detector, which can be placed at any one of five positions along the detector tube.

diffractometer will typically use neutrons with wavelengths in the range 4 to 15 Å and the maximum scattering angles will be small (less than 6°) so that the Q range is of the order of 10^{-3} to 10^{-1} Å^{-1}. The actual range depends on the configuration of the diffractometer which is used. The majority of the SANS experiments which have so far been undertaken on polymer solutions have been carried out using the small-angle diffractometers D11[8] and D17 at the Institut Laue-Langevin, Grenoble, France.[†] A number of other diffractometers have been built or are in the process of being built: at Jülich (West Germany), Harwell (England), Saclay (France), Oak Ridge (USA), the National Bureau of Standards (USA), and the University of Missouri (USA). All these diffractometers are based on similar designs which can be best illustrated by describing D11, the main features of which are shown in Figure 1.

2.1.1. The Small-Angle Diffractometer D11[8]

A beam of long wavelength neutrons from a cold source moves through a curved neutron guide which removes unwanted fast neutrons and γ rays. The beam is then monochromated by passing through a helical slot velocity selector, the speed of which can be varied. For any particular speed only neutrons with the appropriate velocity, i.e., wavelength, can travel through the slots, all others being absorbed. The width of the wavelength distribution depends on the size of the slots. On D11 two selectors are available with resolutions ($\Delta\lambda/\lambda$) of 10% and 50%. As the resolution is relaxed from 10% to 50% the flux increases by a factor of 3. A narrow wavelength dis-

---

† The Institut Laue-Langevin, Grenoble, France, is funded jointly by France, Germany, and the United Kingdom to make available neutron scattering facilities to their scientists who go to ILL to carry out their experiments. Those interested in further information should write to: The Scientific Secretariat, ILL, BP156X Centre de Tri, 38042 Grenoble, France.

tribution is not always necessary, e.g., when measuring radii of gyration, if the shape of the distribution is known, therefore advantage can be taken of this fact if the scattering from the sample is small in order to obtain acceptable statistics in a reasonable length of time.

The monochromated beam then passes through a number of collimators and diaphragms which can be moved into or out of the beam, thus changing the apparent position of the source of neutrons from the sample. On D11 up to 40 m of collimation are available. The neutrons which are scattered by the sample are detected by a position-sensitive detector placed in a 40 m long evacuated tube. The position of the detector can be varied from 1.3 m to 40 m from the sample. The detector is a $^{10}BF_3$ position-sensitive detector with an array of 64×64 1 cm apart electrodes giving 4096 1 cm^2 detector cells. The data are collected from each cell and stored on a dedicated computer.

2.1.2. Experimental Considerations

Although SANS experiments are essentially identical to light-scattering and small-angle x-ray scattering (SAXS) experiments, there are a number of differences which must be remembered. The object of most small-angle scattering experiments is to obtain the scattering due to the polymer alone when it is in the presence of a solvent. The excess scattering of the solution over that of the solvent should give the polymer scattering. However, in SANS there is both incoherent and coherent scattering from the polymer, and because only the coherent scattering contains structural information it is necessary to subtract the incoherent scattering. This is normally achieved by measuring the scattering from a mixture of the solvent plus small molecules, so that the scatterers are randomly distributed and thus will only scatter essentially incoherently. The concentration of these molecules is chosen so that the incoherent scattering from them is equal to that from the polymer. For a hydrogen-containing polymer the incoherent scattering will be significant and has to be subtracted, but if hydrogen is not present then it is small and can often be neglected.

The amount of incoherent scattering also determines the maximum thickness of a sample. It is usual to work with a sample with transmission T less than about 50% so as to optimize the scattered intensity. The optimum thickness is calculated using the total cross-section σ of the solution, which includes both σ_{inc} and σ_{coh}, using the Beer–Lambert law

$$\frac{I'}{I_0} = T = \exp(-n\sigma d) \tag{10}$$

where I_0 and I' are the incident and transmitted intensities, respectively; d is the thickness of the sample and n is the number of scatterers per unit area. For a solution containing mainly hydrogen, σ is dominated by the incoherent cross-section (see Table 1), and the optimum thickness of such a sample is only about 1 to 2 mm. However, a solution containing mainly deuterium or atoms other than hydrogen can be as thick as 5–10 mm. The samples are considerably larger than those used for SAXS.

Usually silica cells are used as containers since silica has a small cross-section (see Table 1) and there is very little small-angle scattering in the Q range used for SANS experiments. The beam size is much larger than for SAXS, of the order of 1–2 cm in diamteer, but it must be remembered that the SAXS flux is much higher than that for SANS. Typical SANS counting times for polymer solutions are between 10 min and 1 hr.

In order to obtain absolute intensities, it is necessary to normalize the data to a standard scatterer and, in this way, take account of a number of factors. As the detector efficiency ε may change across the detector, it must be measured together with the incident intensity I_0 and the solid angle of scatter Ω. It is also necessary to take account of absorption in the sample and this is done by using the transmission T. In order to take account of ε, I_0, and Ω a standard scatterer such as water or vanadium can be used. Any standard which is used to correct for variations in ε must scatter incoherently in the Q range of the experiments, i.e., there must be no angular dependence of the scattering. The absolute intensity can also be normalized using a standard polymer sample of known molecular weight so that the intensity at $Q = 0$ is known (see Section 3.1). Common standards used in SANS are vanadium and water.

2.1.3. Data Analysis

The counts in each cell of the detector are averaged as a function of r, the distance from the center of the beam. Typical data for a solution and a solvent are shown in Figure 2. The scattered intensity I is obtained by subtracting the solvent scattering (plus any incoherent scattering from the polymer) from that of the solution. The scattered intensity is related to the coherent cross-section per scatterer in the following way:

$$I = I_0 \Omega \varepsilon A d T N \left(\frac{d\sigma}{d\Omega} \right)_{\text{coh}} \tag{11}$$

where A is the area of the beam and N is the number of scatterers per unit

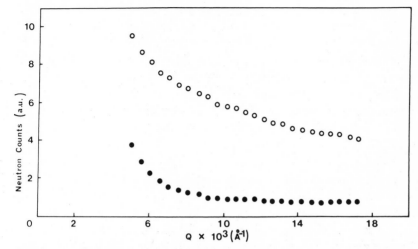

FIGURE 2. Typical small-angle scattering data for a deuterated polymer in a hydrogenous solvent, ○, and the solvent alone, ●.

volume. I_0, Ω, ε, and A are normally assumed to be constant for a series of experiments.

Since the standard scatters incoherently then its scattered intensity I_s is

$$I_s = I_0 \Omega \varepsilon A d T_s N_s \left(\frac{d\sigma_s}{d\Omega} \right)_{\text{inc}} \tag{12}$$

where T_s and N_s are the standard transmission and the number of standard scatterers per unit volume. For an incoherent scatterer the scattered intensity is spread uniformly over 4π sr; therefore Eq. (12) may be approximated to

$$I_s \approx I_0 \Omega \varepsilon A \frac{(1 - T_s)}{4\pi} \tag{13}$$

Substitution for $I_0 \Omega \varepsilon A$ in Eq. (11) gives

$$I = I_s \frac{dT 4\pi}{(1 - T_s)} N \left(\frac{d\sigma}{d\Omega} \right)_{\text{coh}} \tag{14}$$

$$= DN \left(\frac{d\sigma}{d\Omega} \right)_{\text{coh}} \tag{15}$$

where $D = I_s dT 4\pi / (1 - T_s)$. Having normalized the data they can then be analyzed (see Section 3.1).

2.2. High-Resolution Quasielastic Scattering

The 1–10 Å neutrons in the Maxwellian distribution of a thermal reactor have corresponding energies of 80 to 0.8 meV. These energies translate to a frequency range of 2×10^{13} to 2×10^{11} Hz.

If large-scale chain reorganizational motion is to be observed, then the resolution required is at least 10^9 Hz, i.e., better than 1% of the incident neutron energy, even if long-wavelength neutrons from a cold source can be used. Even from high flux reactors this requirement makes unacceptably low the signal-to-noise level on conventional apparatus. Monochromation by crystals or phased choppers followed by energy analysis using either crystal analyzers or time-of-flight techniques[1,2] cannot achieve $\Delta E/E$ much better than about 2%–3%.

Two spectometers, both at the Institut Laue-Langevin at Grenoble, have used new methods to improve resolution in quasielastic scattering and have thus made possible the observation of the internal motion of polymer chains in solution. In each case, energy resolution is gained at the expense of other properties in order to conserve flux.

2.2.1. The Back-Scattering Spectrometer, IN10

The back-scattering spectrometer[9] uses a crystal monochromator set at a 180° Bragg angle, which gives an extremely sharp energy spectrum (~ 1 μeV or 10^9 Hz). A mechanical drive, on this crystal, Doppler-shifts the neutron energies and scans small energy changes on scattering from the sample. The spectrometer is shown schematically in Figure 3. Neutrons arrive in a neutron guide from the reactor and are incident on a monochromator crystal. This is silicon oriented so that the 111 planes Bragg-reflect 6.2 Å neutrons at 180°. Because of this 180° reflection the wavelength spread, which is proportional to $\cot\theta \, d\theta$, is very narrow. This 6.2 Å beam travels back along its path to a graphite crystal and is there reflected at 45° through a chopper (the function of which will be explained later) to a sample. Neutrons scattered by the sample are incident on silicon crystal analyzers oriented with the 111 planes again in back-reflection for 6.2 Å neutrons. The analyzers are on curved mountings focused on detectors just behind the sample position. If the sample scatters inelastically the intensity counted will be diminished since some neutrons no longer reach the analyzers with the necessary 6.2 Å wavelength. However, the monochromator crystal is mounted on a Doppler drive which imparts a small shift in energy to the 6.2 Å reflected neutrons. Only if they lose or gain to the sample the

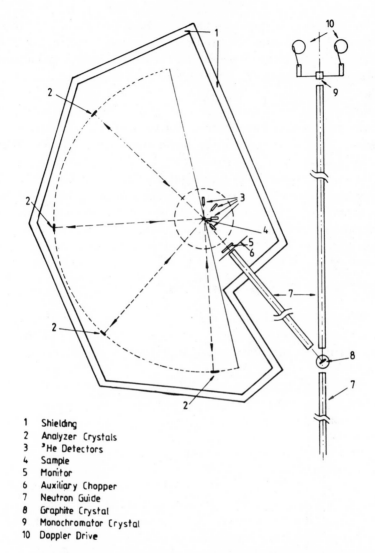

1   Shielding
2   Analyzer Crystals
3   ³He Detectors
4   Sample
5   Monitor
6   Auxiliary Chopper
7   Neutron Guide
8   Graphite Crystal
9   Monochromator Crystal
10 Doppler Drive

FIGURE 3. The back-scattering spectrometer, IN10, at ILL.

energy they have gained or lost at the Doppler drive, so that they again
have exactly 6.2 Å wavelength, will the neutrons be back-reflected at the
analyzers and reach the detectors. The auxiliary chopper coarsely pulses
the neutrons and time-of-flight analysis allows discrimination against
neutrons scattered directly into the detectors. In this way a small energy
scan ($\pm \sim 12$ μeV) can be made with a very sharp (1 μeV) resolution.

As can be seen in this case, resolution has been won at the expense of the width of the observation range. Other limitations concern the Q range and resolution. In order to increase the counting flux, the focusing analyzer crystal banks are made relatively large relaxing the angular resolution and giving a $\Delta Q/Q$ of order 10%. These analyzers can, of course, be masked to give better Q resolution but this is only done for special circumstances because of the severe flux penalties. For similar reasons the lowest angles of scatter used are limited and the smallest Q value is 0.07 Å^{-1}. Even so, there is a danger of some cross-talk at the lowest Q values, and $\Delta Q/Q$ becomes very large. Within these limitations the apparatus has been very successfully used in observation of tunneling splittings of order a few μeV and of quasielastic scattering arising from motion of about 10^9–10^{10} Hz in liquids, macromolecules, and liquid crystals. A typical set of scattering curves for a polymer in solution are shown in Figure 4.

2.2.2. The Spin-Echo Spectrometer, IN11

This technique, originally devised by F. Mezei,[10] uses the precession of the neutron spin in a magnetic guide field as a counter to measure very small changes in neutron velocity. In order to explain the method it is

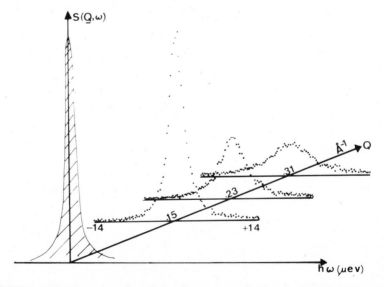

FIGURE 4. $S(\mathbf{Q}, \omega)$ at a number of Q values measured using the IN10 back-scattering spectrometer for a 3% solution of perdeuteropolystyrene in CS_2 at room temperature. The curve at zero Q is the resolution function of the spectrometer.

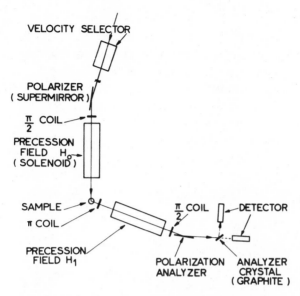

FIGURE 5. The spin-echo spectrometer, IN11, at ILL.

necessary to follow the path of neutrons through the apparatus, which is shown schematically in Figure 5.

The spectrometer[11,12a] has two identical arms on either side of the sample position, each consisting of a length of solenoid providing a magnetic guide field directed along the flight path, and a $\pi/2$ spin turn coil. The incoming neutrons are roughly monochromated using a velocity selector—the wavelength spread $\Delta\lambda/\lambda$ is about 10%. This monochromator is, as will be seen, unnecessary for the *energy* resolution[12b] but dominates the Q resolution. After the polarizer the neutron spins are aligned along the flight path. The $\pi/2$ spin turn coil rotates the neutron spin direction which starts precessing about the guide field. The precession rate of each neutron is the same—given by the Larmor precession frequency ($= \omega_L = 2\mu_n B_0 \hbar$, where μ_n is the magnetic moment and B_0 the guide field strength). However, the number of precessions will be governed by the length of time taken to traverse the guide field, i.e., the neutron velocity. A beam containing a wavelength spread loses its initial phase coherence as it travels through the guide field and, in analogy with NMR, the spins will have fanned out with respect to each other. In the π turn coil, located at the sample, the neutrons make a 180° precession *about a field perpendicular to the guide field* ($H\perp$). This has the effect of reflecting the distribution in the $H\perp$ axis. Those spins that were furthest behind in the fan of precession angles are now furthest ahead and vice versa. Thus, if the length of the second guide field is adjusted to be identical to the first, all the spins will once more be aligned at the second $\pi/2$ turn coil and 100% polarization in the flight path direction, will be observed at the analyzer.

To observe quasielastic scattering the number of precessions in each guide field is set equal for elastic events so that small changes in energy at the sample result in a neutron performing unequal precessions in the two fields and produce a reduction in the polarization recovered at the analyzer. The net polarization $\langle P_z \rangle$ is then given by

$$\langle P_z \rangle = \int_0^\infty I(\lambda) \, d\lambda \int_{-\infty}^{+\infty} P(\lambda, \delta\lambda) \cos \frac{2\pi N_0 \delta\lambda}{\lambda_0} \, d(\delta\lambda) \qquad (16)$$

where $I(\lambda) \, d\lambda$ is the initial wavelength spread, and $P(\lambda)\delta\lambda$ is the probability that a neutron of wavelength λ will be scattered with a wavelength change $\delta\lambda$. Since $\delta\lambda$ can be transformed into an energy change, $\delta E = \hbar\omega$, this is, of course, just the scattering law $S(\mathbf{Q}, \omega)$:

$$\int P(\lambda, \delta\lambda) \, d(\delta\lambda) = \int S(\mathbf{Q}, \omega) \, d\omega \qquad (17)$$

N_0 is the number of precessions made by a neutron of the mean wavelength λ_0. Now since $E = h^2/2m(1/\lambda^2)$

$$\hbar\omega = \delta E = \frac{h^2}{m} \frac{\delta\lambda}{\lambda^3} \qquad \text{or} \qquad \delta\lambda = \frac{m\lambda^3}{2\pi h} \omega \qquad (18)$$

and

$$\langle P_z \rangle = \int_0^\infty I(\lambda) \, d\lambda \int_{-\infty}^\infty S(\mathbf{Q}, \omega) \cos\left\{ \left(\frac{N_0 m\lambda^3}{h\lambda_0} \right) \omega \right\} d\omega \qquad (19)$$

The expression $N_0 m\lambda^3/(h\lambda_0)$ has the dimensions of time, and can be designated $t(\lambda)$. P_z, then, is just the Fourier transform of $S(\mathbf{Q}, \omega)$, i.e., the time correlation function $S(\mathbf{Q}, t)$ is observed directly in spin-echo experiments. The experimental time scale $t(\lambda)$ is governed through N_0 by the strength of the magnetic guide field. For neutrons of 8 Å the time scale covers about two decades from 10^{-9} to 10^{-7} s. For the study of polymer dynamics, direct observation of $S(\mathbf{Q}, t)$ has considerable advantages, as will appear below when analysis of the experimental data is considered. As with the back-scattering spectrometer, the very high resolution (< 0.1 μeV, 10^8 Hz) is won at the expense of Q resolution. In this case, the angular definition of Q is very good, since the beam has to be highly collimated within the guide fields. $\Delta\lambda/\lambda$ is, however, quite broad ($\sim 10\%$) in order to increase

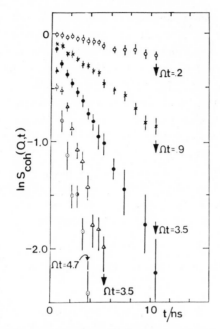

FIGURE 6. Correlation functions at a number of Q values for the same sample as in Figure 4 measured using the spin-echo spectrometer, IN11. Q values: $\bigcirc$, 0.026; $\times$, 0.05; $\bullet$, 0.08; $\triangle$, 0.11; $\odot$, 0.13 Å^{-1}. (From Reference 70.)

flux and this imposes a 10% limit on $\Delta Q/Q$. On the other hand, the tight collimation does allow use of small ($\sim 1°$) scattering angles and the smallest Q available is ~ 0.025 Å^{-1}.

One further important property of the spin-echo technique is its ability to distinguish coherent from incoherent scattering. Most incoherence arises from the neutron spin and, since this machine explicitly follows the spin, it is possible to select only the coherently scattered neutrons. In cases where the coherent and incoherent scattering laws differ, this property is a great advantage. In addition, at low Q, the coherent scattering law has high intensity for macromolecular systems, thus conveniently increasing the signal-to-noise ratio. A typical set of correlation functions for a polymer solution is shown in Figure 6.

2.2.3. Experimental Requirements

Solution samples are held in flat cans perpendicular to the neutron beam which has a diameter of around 3–5 cm, depending on the apparatus and collimation. The sample thickness is conventionally calculated to scatter around 10% of the beam because of multiple scattering considerations. At small Q it is again more usual to take the optimum transmitted scattering intensity which occurs for a sample of 50% transmission [calculated using the Beer–Lambert law, Eq. (10)]. Thin-walled aluminum cans are used at higher Q values but because of the small-angle scattering from most aluminum, quartz or niobium is used at small angles. Materials must be nonmagnetic for the spin-echo spectrometer. Sets of data, such as those shown in Figs. 4 and 6, took about 24 hr to obtain on the high flux reactor at the ILL. The scattering from an empty container is subtracted from the data. The resolution function is measured by scattering from an elastic scatterer—either vanadium, for high-Q incoherent work, or a glassy polymer containing labeled chains, for low-Q coherent experiments.

2.2.4. Analysis of Quasielastic Scattering from Macromolecules in Solution

In studying dynamics of macromolecules in solution, the quasielastic scattering is usually compared with calculations based on models, for example,[13] the scattering law for a chain undergoing Rouse or Zimm internal motion (see Chapter 1). This comparison is complicated by the fact that observation is frequently made close to the resolution limits of even the highest-resolution spectrometers. The energy spread in the initial beam is convoluted with that introduced by the scattering process, and may well be of the same order of magnitude. Deconvolution is unreliable in these circumstances and, in general, a procedure is adopted where the proposed model is convoluted with the observed instrumental resolution (usually obtained as scattering from a purely elastic scatterer, e.g., vanadium) and an iterative fit made to the experimental data. The information extracted is therefore strongly dependent on the initial choice of model. A more demanding test of a model is to fit it to data from the same sample at two very different experimental resolutions.[13]

The Fourier transform of a convolution is a product, i.e.,

$$FT[S(\mathbf{Q}, \omega) * I(\omega)] = FT[S(\mathbf{Q}, \omega)] \times FT[I(\omega)] \tag{20}$$

It is particularly simple, therefore, to remove the instrumental resolution from data obtained on the spin-echo spectrometer. The data are divided by the spectrum from an elastically scattering sample, and the resulting $S(\mathbf{Q}, t)$ is essentially model independent.

3. THE CONFORMATION OF MACROMOLECULES IN SOLUTION

3.1. Introduction

Small-angle neutron scattering has changed from being a little-used, unknown technique to a major experimental method for the study of polymer conformations in little over ten years. During this period SANS has not only shed a great deal of light on bulk polymer conformations but also on, for example, the excluded-volume behavior of semidilute and concentrated solutions, conformations of block copolymers, and partially labeled polymers and polyelectrolytes in solution. Unlike SAXS or light scattering, it is possible to label parts of the chains selectively with deuterium without changing the properties of the system. This major advantage of SANS has also encouraged an upsurge in theoretical interest in the conformations and excluded-volume behavior of polymers, which has added enormously to the knowledge and understanding of polymers. The scope for research which SANS offers in the field of solution properties of macromolecules is only just beginning to be exploited.

3.1.1. Basic Small-Angle Scattering Theory

The theory described in this section is not intended to be an exhaustive treatment of the subject; rather, it is hoped, it will give a basis on which to understand the experimental results. Although the theory described here is oriented toward SANS, it is also applicable, with modifications where necessary, to any low-Q technique, e.g., SAXS and light scattering.

The differential cross-section, given by Eq. (2), is composed of two components, an incoherent cross-section $(d\sigma/d\Omega)_{\text{inc}}$ and a coherent cross-section $(d\sigma/d\Omega)_{\text{coh}}$. It is the coherent cross-section of a macromolecular solution which contains the structural information. In the following discussion of the theory it is assumed that the incoherent scattering from the components of the solution has been subtracted leaving only the coherent scattering.

Consider a solution of polymer 1 and solvent 0; then the coherent cross-section for the solution can be written[14]

$$\left(\frac{d\sigma}{d\Omega}\right)_{\text{coh}} = \left\langle b_1{}^* \sum_i \exp(i\mathbf{Q} \cdot \mathbf{R}_{1i}) + b_0{}^* \sum_j \exp(i\mathbf{Q} \cdot \mathbf{R}_{2j}) \right\rangle^2 \quad (21)$$

where $b_1{}^*$ and $b_0{}^*$ are the total coherent scattering lengths per unit volume of the monomer and solvent molecules, respectively. $b_1{}^* = N_A \sum_n b_{1,n}/V_1$ and $b_0{}^* = N_A \sum_n b_{0,n}/V_0$, where $b_{1,n}$ and $b_{0,n}$ are the atomic scattering lengths of the atoms comprising the monomer and solvent molecules. V_1 and V_0 are partial molar volumes of a monomer unit and solvent molecule, respectively. N_A is Avogadro's number. Expanding Eq. (21) gives

$$\left(\frac{d\sigma}{d\Omega}\right)_{\text{coh}} = b_1{}^{*2}\left\langle \sum_i \exp(i\mathbf{Q}\cdot\mathbf{R}_{1i})\right\rangle^2 + b_0{}^{*2}\left\langle \sum_j \exp(i\mathbf{Q}\cdot\mathbf{R}_{2j})\right\rangle^2$$

$$+ b_1{}^*b_0{}^*\left\langle \sum_i \exp(i\mathbf{Q}\cdot\mathbf{R}_{1i})\right\rangle\left\langle \sum_j \exp(i\mathbf{Q}\cdot\mathbf{R}_{2j})\right\rangle$$

$$+ b_0{}^*b_1{}^*\left\langle \sum_j \exp(i\mathbf{Q}\cdot\mathbf{R}_{2j})\right\rangle\left\langle \sum_i \exp(i\mathbf{Q}\cdot\mathbf{R}_{1i})\right\rangle \qquad (22a)$$

which can be simplified to give

$$\left(\frac{d\sigma}{d\Omega}\right)_{\text{coh}} = b_1{}^{*2}S_{11}(Q) + b_0{}^{*2}S_{00}(Q) + b_1{}^*b_0{}^*S_{10}(Q) + b_0{}^*b_1{}^*S_{01}(Q) \quad (22b)$$

where

$$S_{kk}(Q) = \left\langle \sum_i \exp(i\mathbf{Q}\cdot\mathbf{R}_{ki})\right\rangle^2$$

$$\text{and} \quad S_{nk}(Q) = \left\langle \sum_j \exp(i\mathbf{Q}\cdot\mathbf{R}_{nj})\right\rangle\left\langle \sum_i \exp(i\mathbf{Q}\cdot\mathbf{R}_{ki})\right\rangle \qquad (22c)$$

$S_{nk}(Q)$ are scattering functions and are directly related to the equivalent density–density correlation functions. If there are no density fluctuations in the solution, i.e., the solution is incompressible, then

$$\sum_n^P S_{nk}(Q) = 0 \qquad (23a)$$

or, in the case being considered here,

$$S_{00}(Q) + S_{10}(Q) = 0, \qquad S_{01}(Q) + S_{11}(Q) = 0$$

thus

$$S_{00}(Q) = -S_{10}(Q) \qquad (23b)$$

$$S_{11}(Q) = -S_{01}(Q) \qquad (23c)$$

therefore Eq. (22b) becomes

$$\left(\frac{d\sigma}{d\Omega}\right)_{\text{coh}} = (b_1{}^* - b_0{}^*)[b_1{}^*S_{11}(Q) - b_0{}^*S_{00}(Q)] \qquad (24)$$

but since

$$S_{01}(Q) = S_{10}(Q)$$

then from Eqs. (23b) and (23c) it can be seen that

$$S_{11}(Q) = S_{00}(Q)$$

and therefore Eq. (24) becomes

$$\left(\frac{d\sigma}{d\Omega}\right)_{\text{coh}} = (b_1{}^* - b_0{}^*)^2 S_{11}(Q) \qquad (25a)$$

$$= K_1{}^2 S_{11}(Q) \qquad (25b)$$

where $(b_1{}^* - b_0{}^*) = K_1$, the contrast factor between polymer and solvent. K_1 determines whether or not any scattering will be observed, it is equivalent to the refractive index increment in light-scattering and the electron density difference in SAXS.

Equation (25) can be generalized for any multicomponent solution composed of $p + 1$ components $(0, 1, 2, \ldots, p)$ to give[15]

$$\left(\frac{d\sigma}{d\Omega}\right)_{\text{coh}} = \sum_{n=1}^{P} \sum_{k=1}^{P} (b_n{}^* - b_0{}^*)(b_k{}^* - b_0{}^*) S_{nk}(Q) \qquad (26a)$$

or

$$\left(\frac{d\sigma}{d\Omega}\right)_{\text{coh}} = \sum_{n=1}^{P} \sum_{k=1}^{P} K_n K_k S_{nk}(Q) \qquad (26b)$$

where K_n (or K_k) is the contrast factor between component n (or k) and component 0. The $S_{nk}(Q)$'s are functions of $P_{1,k}(Q)$, the intramolecular or single chain structure factor, and $P_{2,nk}(Q)$, the intermolecular structure factor and can be written

$$S_{kk}(Q) = N_k Z_k{}^2 P_{1,k}(Q) + N_k{}^2 Z_k{}^2 P_{2,kk}(Q)$$

$$S_{nk}(Q) = N_n N_k Z_n Z_k P_{2,nk}(Q) \qquad (27)$$

where there are N_n chains of polymer n, each chain of which is made up of Z_n monomers. $N_n = c_n N_A \mid M_{n,w}$ and $Z_n = M_{n,w} \mid m_n$, where c_n, $M_{n,w}$ and m_n are the concentration in g/ml, the weight average molecular weight and the monomer molecular weight of polymer n, respectively. To calculate the scattered intensity, the contrast factor per monomer K^* is required

(K_n is the contrast factor per unit volume); therefore

$$K_n{}^* = \frac{V_n}{N_A}(b_n{}^* - b_0{}^*)$$

$$= \frac{m_n}{\varrho_n N_A} K_n$$

where ϱ_n is the density of component n. Substituting K^* for K and using Eq. (27) for $S_{11}(Q)$ in Eq. (25), one obtains

$$\left(\frac{d\sigma}{d\Omega}\right)_{\text{coh}} = K_1{}^{*2}[N_1 Z_1{}^2 P_{1,1}(Q) + N_1{}^2 Z_1{}^2 P_{2,11}(Q)] \tag{28a}$$

and substituting for N_1 and Z_1 this becomes

$$\left(\frac{d\sigma}{d\Omega}\right)_{\text{coh}} = \frac{K_1{}^2}{\varrho_1{}^2 N_A}[c_1 M_{1,w} P_{1,1}(Q) + c_1{}^2 N_A P_{2,11}(Q)] \tag{28b}$$

Using Eq. (15), Eq. (28) can now be written in terms of intensity remembering that $(d\sigma/d\Omega)_{\text{coh}}$ in Eq. (28) is the cross-section of the total number of scatterers. The scattered intensity from a polymer solution is

$$I = R[c_1 M_{1,w} P_{1,1}(Q) - 2c_1{}^2 M_{1,w}^2 A_2(Q)] \tag{29}$$

where $R = K_1{}^2 D/(\varrho_1{}^2 N_A)$ and $A_2(Q) = -N_A P_{2,11}(Q)/(M_{1,w}^2 2)$. Note that no assumptions have been made about the concentration or Q range when deriving Eq. (29). In the dilute regime Eq. (29) inverts to give the familiar Zimm equation

$$\frac{Rc_1}{I} = \frac{1}{M_{1,w} P_{1,1}(Q)} + 2A_2(Q)c_1 \tag{30}$$

where A_2 is the second virial coefficient. $P_1(Q)$ has been calculated for a Gaussian chain by Debye[17]:

$$P_1(Q) = (2/x^2)[\exp(-x) - 1 + x] \tag{31}$$

where $x^2 = Q^2 R_g{}^2$ and $R_g{}^2$ is the mean square radius of gyration of the polymer chain. It is useful to consider the limiting values of $P_1(Q)$ using Debye's expression:

(1) In the range $QR_g < 1$, the Guinier range, Eq. (30) becomes

$$\frac{Rc_1}{I} = \frac{1}{M_{1,w}}\left(1 + \frac{Q^2 R_g{}^2}{3}\right) + 2A_2(Q)c_1 \tag{32a}$$

and (2) in the range $R_g^{-1} < Q < l^{-1}$, where l is the persistence length, Eq. (30) becomes

$$\frac{Rc_1}{I} = \frac{Q^2 R_g^2}{2M_{1,w}} + 2A_2(Q)c_1 \qquad (32b)$$

Using Eq. (32a) it is possible to obtain $M_{1,w}$, R_g, and A_2 by extrapolating Rc_1/I to $c_1 = 0$ and $Q = 0$ and measuring the slopes and intercepts of the appropriate lines.

3.2. Applications to Macromolecules in Solution

3.2.1. Dilute Solution

In dilute solution, the H–D labeling technique does not, in general, give neutron scattering an advantage over light or x-ray scattering, since it is not necessary to distinguish molecules from otherwise identical surrounding molecules. While deuteration of polymer or solvent will usually be used to increase contrast and therefore signal-to-noise ratio, the justification for use of neutrons must be in the domain of Q it is necessary to explore.

The conformation of macromolecules in solution has been extensively studied by light scattering for molecular dimensions greater than about 200 Å. For smaller molecules, x-ray techniques have, in general, proved adequate. There have, however, been limited results published using neutron scattering, which show that if the restricted availability is ignored it is a much easier technique to use for studying smaller macromolecules in solution. This is both because use of deuterated solvents (commonly available for NMR) increases the contrast and because the Q range defining the distance scale is larger than that for SAXS.

For observation of very small molecules, the increased contrast can be important since signal intensity is proportional to M_w.

The scattering from molecules in solution in the range $QR_g < 1$ is usually displayed in the form of a Zimm plot[16] [see Eq. (32)].

For θ solvents (cf. W. Forsman, Chapter 1) $A_2 = 0$ and extrapolation to zero concentration is unnecessary to obtain $P_1(Q)$ and hence R_g. Ballard et al.[19] observed low-molecular-weight polystyrene under θ conditions in deuterated cyclohexane. They showed that for values of M_w below about 3000 the dimensions are no longer given by the simple expression for a Gaussian random coil,

$$R_g = AM^{1/2} \qquad \text{where } A = 0.275 \text{ Å for polystyrene}$$

Chain stiffness gives rise to a more linear conformation and an effective reduction in the radius of gyration, which in this case, fitted well to calculations based on the "wormlike chain" model of Kratky and Porod.[20]

Higgins et al.[21] used SANS to observe the difference, in conformation and dimensions, of cyclic and linear poly(dimethyl siloxane) molecules of corresponding molecular weights. Here both the increased contrast and the wide Q range were important. Figure 7 shows the $c/I(Q)$ plots for a ring and a chain sample of $M_w \sim 20{,}000$ in deuterated benzene. Both curves show a linear region at low Q but the upward deviations are apparent at

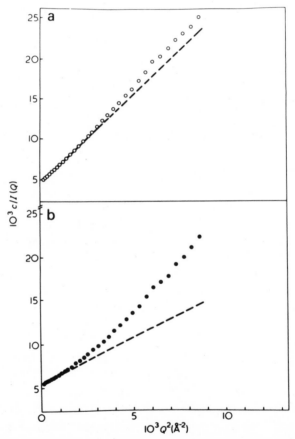

FIGURE 7. Plots of inverse neutron scattering intensities $c/I(Q)$ against Q^2 for (a) a linear poly(dimethyl siloxane) fraction with z-average molecular weight $\bar{M}_z = 20{,}880$ and concentration $c = 0.0362$ g cm^{-3} in benzene-d_6 at 292 °K and (b) a cyclic poly(dimethyl siloxane) fraction with $\bar{M}_z = 20{,}210$ and $c = 0.0349$ g cm^{-3} under the same conditions. The broken lines are extrapolations of the linear regions at low Q. (From Reference 21.)

much lower Q for the cyclic sample. The deviations from linearity of a $c/I(Q)$ plot occur when the Guinier limit, $QR_g < 1$, is exceeded and internal conformation of the coil is explored. This local conformation is ultimately unaffected by the cyclic nature of the ring molecules, so that at high Q the scattering from the rings and chains is identical. However, the overall dimensions are affected and these authors showed that the ratio of $R_g{}^2$ for a chain to that of a ring molecule was approximately 2. Thus, at low Q, the slope of the Zimm plot for a cyclic molecule is much less than that for the corresponding chain and it is this effect that is responsible for the pronounced upward curvature for the ring molecule in Figure 7. Experimentally, it was necessary to explore a wide Q range to be certain of observing the true low-Q limit for the ring molecules. Outside the Guinier range a number of regions of behavior can be distinguished. Equation (32b) details the scattering behavior in the range $R_g{}^{-1} < Q < l^{-1}$.

For $l^{-1} \leq Q \leq r^{-1}$, where r is the radius of the chain cross-section, the scattering appears as that from a rigid rod of length[22] L

$$I(Q) = Rc \frac{M_w}{L} \pi Q^{-1} \tag{33}$$

Gupta et al.[23] explored these higher-Q regions for solutions of cellulose tricarbanilate in dioxane and determined the persistance length l, from the transition between the coil and rigid rod regions, as a function of temperature. Again, they concluded that the good contrast and wide Q range made the neutron experiments faster and easier than using x-rays.

3.2.2. Semidilute and Concentrated Solutions

3.2.2a. Theoretical Background. As the concentration of polymer in a good solvent increases from infinite dilution to the bulk polymer density, the excluded volume varies from a positive value to zero. The extremes of the concentration range have been well dealt with theoretically, but not the intermediate concentrations. During the past few years a great deal of theoretical interest has been shown in developing a theory which can explain the excluded-volume behavior of polymer chains at any concentration. This renewed interest can be traced back to a paper by Edwards[24] in which it was argued that polymer solutions could be classified into three types. These regimes of behavior are characterized in terms of the step length l, the total length of the chain R, the excluded volume β, and the monomer density ϱ.

Using mean field theory, Edwards defined the three regions of concentration:

(1) the dilute regime where the chains are isolated;
(2) the intermediate concentration regime, where the chains overlap but the segment density is not large; and
(3) the dense regime, where every segment is in contact with many others.

Edwards[24] for the first time introduced the concept of a Debye–Hückel type of screening length (ξ) into polymer theory. In a dilute solution the polymer chains are well separated and the segmental interactions which they experience are intramolecular. It is these interactions which give rise to the excluded-volume effect. As the concentration increases, the chains overlap giving rise to intermolecular interactions which screen interactions between distant parts of the same chains, thus decreasing the excluded volume. The screening length is a measure of this weakening of the excluded-volume effect and is the average distance along a chain between intermolecular contacts. It can be thought of as the distance beyond which there are no excluded-volume interactions between parts of the same chain. It is this concept of screening length which is the basis of the two main theoretical approaches to the problem of polymer solution properties, the mean field theory (MF) of Flory[25] and Edwards[24] and the renormalization group–scaling law (SL) approach of de Gennes[26] and des Cloiseaux[27].

de Gennes[26] and des Cloiseaux[27] have used an analogy between polymer solutions and magnetic critical point phenomena to apply renormalization group theory and scaling arguments to the theory of polymer solutions. de Gennes pointed out that the Flory θ temperature of a polymer solution could be considered to be the same as a tricritical point in magnetic systems.[28] Using this concept and SL arguments, Daoud and Jannink[29] were able to calculate the way in which the properties of the solution vary in different concentration regions.

In order to use scaling arguments it is necessary to split up the chain into a number of sections or "blobs." Within each blob the chain exhibits excluded-volume behavior, but the chain as a whole is Gaussian and the effective interaction between the blobs is weak. The length of these blobs is equivalent to the screening length ξ.

A distinction was drawn between the tricritical and critical regions and a temperature-concentration diagram was constructed with lines indicating the cross-over points between different regions. This so-called "phase" diagram indicated a number of regions of behavior (Figure 8):

I', the dilute tricritical θ region;

I, the dilute critical region (good solvent);

II, the semidilute critical region (good solvent);

III, the semidilute tricritical θ region; and

IV, the coexistence region.

The phase boundaries in Figure 8 are defined as follows:

(a) The boundary between II and III is c^{**}:

$$c^{**} \sim \tau \qquad [\tau = (T - \theta)/\theta]$$

(b) The boundary between I and I' is

$$\tau \simeq Z^{-1/2} \qquad (Z = \text{degree of polymerization})$$

(c) The boundary between I and II is c^*, the overlap concentration:

$$c^* \sim Z/S^3 \qquad (S = \text{end-to-end distance})$$

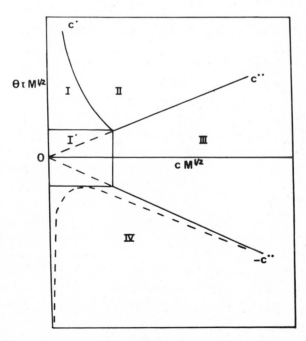

FIGURE 8. Temperature-concentration diagram for a polymer solution as predicted by renormalization group theory. The regions are explained in the text.

TABLE 2. Theoretical Predictions for R_g and ξ Calculated Using Either the Mean-Field Theory (MF) or the Scaling Law (SL) Approach

Region	R_g^2 (SL)	R_g^2 (MF)	ξ^2 (SL)	ξ^2 (MF)
I'	Z	Z		
I	$Z^{6/5}\beta^{2/5}$	$Z^{6/5}\beta^{2/5}$	—	—
II	$Z\varrho^{-1/4}\beta^{1/4}$	$Z\varrho^{-1/5}\beta^{1/4}$	$\varrho^{-3/2}\beta^{-1/2}$	$\varrho^{-6/5}\beta^{-4/5}$
II A	—	$R_0^2[1 + K\beta^{1/2}\varrho^{-1/2}]^a$	—	$\varrho^{-1}\beta^{-1}$
III	Z	Z	ϱ^{-2}	—

^{a} R_0 is the unperturbed radius of gyration.

These "phase" boundaries are not considered to be sharp but simply the mean point between two asymptotic laws. In each region, the radius of gyration, R_g, and the screening length, ξ, can be written in terms of the reduced temperature, τ, the concentration, c, and the degree of polymerization, Z (Table 2).

Previous to the development of the SL theory, Edwards[30] had calculated the behavior of R_g and ξ in a "strong" solution. The results of this calculation are shown in Table 2 as region IIA. Recently Edwards and Jeffers[31] have formulated an expression for the behavior of polymer chains in solution which can be extrapolated from infinitely dilute solution to the bulk density. This formula predicts almost identical behavior to that obtained using the SL approach. Differences between the theories are found in the predictions for region II. Edwards and Jeffers predict two regions: II, the semidilute, where differentiation between the two sets of exponents would be difficult experimentally; and IIA, the semiconcentrated region, which is not predicted by SL theory.

3.2.2b. Experimental Results. One of the major contributing factors to the increase in theoretical interest in concentrated polymer solutions has been the advent of SANS. SANS has enabled measurements of single-chain correlations to be made on concentrated solutions by labeling some of the chains. It has been possible to test the theories experimentally using SANS by making measurements of both single-chain and multichain correlations. Multichain correlation measurements are made on solutions in which the chains overlap, i.e., $c > c^*$, the overlap concentration, and all the chains are identical. On the other hand, single-chain measurements are made on mixtures of protonated polymer (concentration c_p) and deuterated polymer (concentration c_D) such that $c(= c_p + c_D) > c^*$, but $c_D < c^*$. Therefore, the labeled chains are isolated, dilute relative to each other.

Inherent in the use of this labelling technique is the assumption that protonated and deuterated chains are thermodynamically identical. In fact, there can be quite a difference in their thermodynamic properties. Strazielle and Benoit[32] measured the θ temperature of polystyrene (PSH) and poly-(deuterostyrene) (PSD) in cyclohexane and deuterated cyclohexane and discovered that the θ temperatures varied from 30°C (PSD in C_6H_{12}) to 40°C (PSH in C_6H_{12}). Although there is a variation in the θ temperatures, the difference in the interaction parameters between PSH and PSD in C_6H_{12} is very small. Therefore, Benoit and Strazielle[32] argue that it is probable that any enthalpy of mixing between the polymers in the absence of solvent will also be very small. When investigating concentrated solutions, the assumption is made that the deuterated and protonated chains behave in the same way with the proviso that when measuring the temperature dependence of single-chain correlations (R_g and ξ) the value of the θ temperature which may be calculated from the data, or used in subsequent analysis, is that of the labeled material (see, for example, References 33 and 34).

Farnoux and coworkers[35] have investigated the momentum transfer dependence of the single-chain and multichain scattering amplitudes $S_1(Q)$ and $S(Q)$, respectively, of a polymer in a good solvent. The system they used was polystyrene in carbon disulfide. The measurements were made in the intermediate Q range, $(N_s^{1/2}l)^{-1} < Q < l^{-1}$, where N_s is the number of steps of length l in the chain. The theoretical expression for $S(Q)$ is

$$S(Q) = \frac{Ac}{(Q^2 + \xi^{-1})} \tag{34}$$

There is no analytical expression for $S_1(Q)$, so the form that was used was

$$S_1(Q) = \frac{Bc}{Q^{1/\nu} + O(1/N_s l^{1/\nu})} \tag{35}$$

A and B are constants and ν is the excluded-volume exponent. At the two extremes of the concentration range the values of ν are known. In dilute solution, where there are excluded-volume effects $\nu = 3/5$, and in the bulk, where these effects go to zero $\nu = 1/2$. According to both the MF and SL theories, as the chains overlap the distance over which excluded-volume effects are important should decrease as these effects are screened out.

Data were collected from two semidilute solutions, a concentrated solution, and a bulk sample and plotted as $I(Q)^{-1}$ vs. Q^2 and vs. $Q^{5/3}$. As the concentration increased, the linear dependence of $[S_1(Q)]^{-1}$ on $Q^{5/3}$ covered less of the Q range. Eventually $[S_1(Q)]^{-1}$ became proportional to Q^2 as the concentration increased to the bulk density. The multichain term $[S(Q)]^{-1}$ was proportional to Q^2 for both the semidilute solutions. $[S_1(Q)]^{-1}$ was found to be proportional to either Q^2 or $Q^{5/3}$ depending upon the Q range and the concentration. The conclusion was drawn that at a given concentration ($c > c^*$) there is a value of $Q = Q^*$ which separates Q space into two regions. For $Q < Q^*$, there is a strong screening effect and $\nu = 1/2$, and for $Q > Q^*$, the chains exhibit excluded-volume effects and $\nu = 3/5$. The value of Q^* moves from R^{-1} to l^{-1} as the concentration increases from the dilute region to the bulk state. This cross-over point in reciprocal space Q^* is related to the screening length ξ.

The single-chain cross-over Q^* can be used to measure ξ but is not as accurate as the multichain method as the exact cross-over point is difficult to pinpoint. Equation (34) is normally employed to calculate ξ using the type of plot shown in Figure 9. ξ can be calculated from either the slope or the intercept of the data.

This concept of the spatial cross-over has been further explored and enlarged upon[36] and Q^* has been measured as a function of c and τ (the reduced temperature). Using scaling law arguments the screening length is related to a critical number of segments which varies with temperature or concentration depending upon whether the chains are in a good solvent in (1) the zero concentration limit or (2) the semidilute region, respectively. The screening length in the semidilute region is expressed as

$$\xi_c^2 = n_{cc}^2 l^2 \tag{36}$$

where $\nu = 3/5$ and $n_{cc} \sim c^{-5/4}$; therefore

$$Q^* \sim c^{3/4} \tag{37}$$

In the dilute region, $\xi_D^2 = n_c l^2$ and $n_c = \tau^{-2}$; therefore

$$Q^* = \tau \tag{38}$$

Both expressions were found to be a reasonable fit to the data in the appropriate regime.

The predicted laws relating to the semidilute region have been investigated in two papers[37,33] which measured the concentration and temperature dependences of R_g and ξ. Daoud and co-workers[37] used polystyrene (M_w

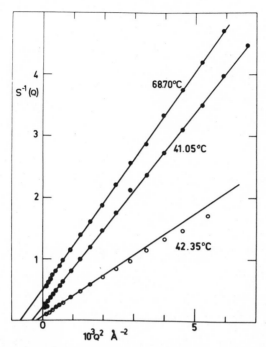

FIGURE 9. Inverse of the scattering cross section $S(Q)$ plotted as a function of Q^2, at two concentrations 0.092 g cm^{-3} ($\bullet$) and 0.058 g cm^{-3} ($\bigcirc$). For clarity the values of $S^{-1}(Q)$ for the lowest concentration are divided by a factor of 5. The departure from linearity of this last curve is due to the cross over from Q^2 to $Q^{5/3}$ behavior as explained in the text. [Reprinted with permission from J. P. Cotton, M. Nierlich, F. Boué, M. Daoud, B. Farnoux, G. Jannink, R. Duplessix, and C. Picot, *J. Chem. Phys.* **65**, 1101 (1976).]

= 114,000 and 500,000) in a good solvent CS$_2$ to study the concentration dependence. Over a range of concentrations up to the bulk, it was found that the SL predictions for region II fitted the data. The temperature dependences of R_g and ξ were measured using polystyrene ($M_w = 160,000$ and $c = 0.15$ g/cm^3) in cyclohexane. Cotton *et al.*[33] constructed a phase diagram from the temperature cross-overs. Figure 10 shows the cross-over as T is increased from θ through the tricritical semidilute region (III) to the critical semidilute region (II). From this and other data, a phase diagram was constructed (Figure 11). The SL predictions were further supported by other measurements on the same system[38] polystyrene/cyclohexane ($M_w = 75,000$).

The semidilute measurements discussed above were all made near the upper critical solution temperature (UCST) and all the theories are based

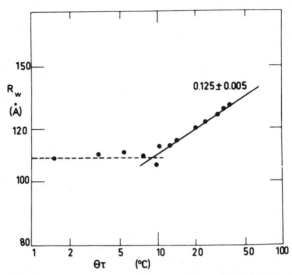

FIGURE 10. Log–log plot of the weight average radius of gyration R_w as a function of $\theta\tau$, showing the cross over from dilute (I') to semidilute (II) behavior. The slope of 0.125 ± 0.05 is in good agreement with the predictions of Daoud and Jannink.[29] The horizontal dashed line gives the value of R_w in θ conditions. [Reprinted with permission from J. P. Cotton, M. Nierlich, F. Boué, M. Daoud, B. Farnoux, G. Jannink, R. Duplessix, and C. Picot, *J. Chem. Phys.* **65**, 1101 (1976).]

on θ being the UCST. Richards and co-workers[34] have extended the temperature range of their measurements toward the lower critical solution temperature (LCST). Close to the LCST there is evidence of a cross-over into a tricritical region (Figure 12). Although the behavior is similar to that close to the UCST the slope of the temperature dependent part is ~ 0.52, twice that predicted by either theory near the UCST. A phase diagram was constructed about the LCST which mirrors the normal diagram based on the UCST.

Edwards and Jeffers[31] predict region IIA (semiconcentrated) and measurements on solutions with concentrations well above c^* have found evidence to support this prediction.[34] Figure 13 shows R_g^2 plotted as a function of $(T - \theta)^{1/2}$ for a 0.47 g/cm³ solution of polystyrene ($M_w = 75,000$) in cyclohexane. The data not only vary as $(T - \theta)^{1/2}$ above θ but also down to 20°C below θ. There would seem to be no evidence of a tricritical or theta region near $T = \theta$. Measurements of ξ also support this view as $\xi^{-2} \propto (T - \theta)$ above and below θ. At high concentrations, above ~ 0.8 g/cm³, R_g is no longer proportional to $c^{-1/2}$ but becomes independent of c and equal to the value of R_g in the bulk.

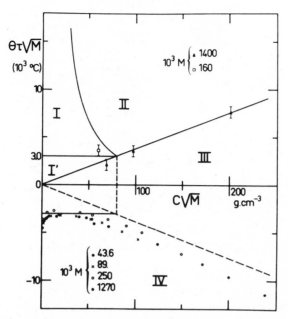

FIGURE 11. The temperature-concentration diagram constructed from the neutron scattering measurements of the temperature dependence of the radius of gyration (□) and the screening lengths (▲). The data points in region IV are phase separation data for polystyrenes of different molecular weights in cyclohexane. [Reprinted from J. P. Cotton, M. Nierlich, F. Boué, M. Daoud, B. Farnoux, G. Jannink, R. Duplessix, and C. Picot, *J. Chem. Phys.* **65**, 1101 (1976).]

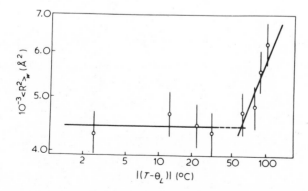

FIGURE 12. Log–log plot of R_w^2 as a function of $|(T - \theta_L)|$, for polystyrene in cyclohexane, $\theta_L = 213\,°C$. (From Reference 34b.)

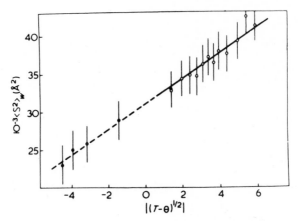

FIGURE 13. Plot of the mean-square end-to-end distance s_w^2 as a function of $|(T - \theta)^{1/2}|$, for $32° \leq T \leq 65\,°C$, $\bigcirc$, and for $10° \leq T \leq 32\,°C$, $\bullet$. (From Reference 39.)

As the temperature is increased from the UCST to the LCST the value of R_g in a semiconcentrated solution goes through a maximum (Figure 14). The data on either side of this maximum fit the type of temperature dependence near the UCST predicted by Edwards,[30] i.e., $\propto (T - \theta)^{1/2}$. Using a corresponding states expansion of the excluded volume, an attempt was made to fit the data. Although the general shape of the curve can be predicted, the absolute numbers are considerably different (Figure 14).

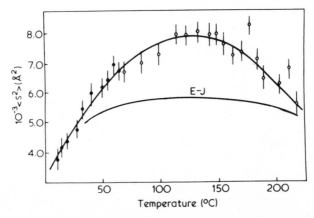

FIGURE 14. Values of R_w^2 as a function of T for polystyrene in cyclohexane (0.477 g cm^{-3} at $25\,°C$, $M_w = 7.57 \times 10^4$). The theoretical curve (E–J) was calculated using the equation for region IIA (Table 2) and a theoretical value for the polymer–solvent interaction parameter from Edwards and Jeffers.[31] (From Reference 34c.)

Although the SL theory predicts the correct behavior in the semi-dilute regime, it does not predict the semiconcentrated regime. It would seem from the SANS data collected so far, that a new phase diagram is required which takes into account the semiconcentrated regime. Recent calculations by Schaefer and coworkers[39] on the dynamics of semiflexible polymers would seem to support this view. They suggest that there are MF regions and SL regions the sizes of which depend upon the quality of the solvent.

3.2.3. Copolymers and Partially Labeled Chains

If a block copolymer, with incompatible blocks, is dissolved in a solvent the repulsive interactions may drive the molecules to form segregated regions even in solution. This segregation can be studied by light scattering if the solvent can be made isorefractive with each component in turn. Unfortunately, this involves changing the chemical nature of the solvent and thus the solvent–polymer interactions. H–D exchange offers a much less violent way of changing the contrast between the polymer components and the solvent, and allows the dimensions of each component to be determined in essentially the same solvent (however, note the caveat in Section 3.2.2 about the effect of H–D substitution on θ temperatures). Han and Mozer[40] used this technique to observe the dimensions of a poly-styrene–polymethylmethacrylate diblock copolymer in toluene. The poly-styrene component was deuterated so that in normal toluene only this block was visible and in perdeutero toluene, the scattering from the PMMA block was dominant. The R_g values obtained for the two blocks indicated that the polystyrene was in a more expanded state than the polymethyl-methacrylate, a result which could possibly be explained by the superior solvent power of toluene for polystyrene.

Contrast matching one component of the copolymer is an extreme case of a more general expression relating the apparent radius of gyration of a multicomponent molecule to the radii of gyration of the components.[41] For a two-component system

$$R_g^2(\text{apparent}) = YR_g^2(1) + (1 - Y)R_g^2(2) + Y(1 - Y)G_{12}^2 \quad (39)$$

where $R_g(1)$ and $R_g(2)$ are the radii of gyration of the components 1 and 2 and G_{12} is the distance between their geometric centers. Y is a mean contrast defined as $Y = K_1/(K_1 + K_2)$ (see Section 3.1). Attempts to observe the dependence of $R_g(\text{apparent})$ on Y, using light scattering, failed

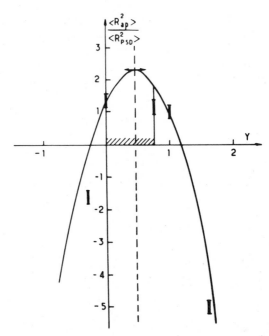

FIGURE 15. Parabolic dependence of apparent $R_g{}^2$ on contrast factor Y for a PSD–PSH diblock copolymer in a mixed C_6H_{12}/C_6D_{12} solvent of varying compositions at 35 °C. [Reprinted with permission from M. Duval, R. Duplessix, C. Picot, D. Decker, H. Benoit, J. P. Cotton, R. Ober, G. Jannink, and B. Farnoux, *J. Polym. Sci.*, Part B, **14**, 588 (1976).]

because changing the refractive index difference necessarily changed the chemical nature of the solvent.

Using mixtures of deuterated and hydrogenous solvent it is possible to scan Y almost continuously without changing the chemical nature of the solvent. R_g (apparent) should then show a parabolic variation with Y which allows determination of $R_g(1)$, $R_g(2)$, and G_{12}. Duval *et al.*[42] demonstrated the parabola (shown in Figure 15) for a PSD–PSH diblock copolymer in a mixed C_6H_{12}/C_6D_{12} solvent of varying compositions at the theta temperature ($\sim$35 °C). The values of R_g obtained for the two blocks at $Y = 0$ and 1, respectively, are rather larger than would be obtained for homopolymers of corresponding length. However, it has been shown[43] that polydispersity strongly affects the shape of the parabola and this probably explains the discrepancy. The distance G_{12} between centers agreed well with the value, given by $G_{12}^2 = 2[R_g(1)^2 + R_g(2)^2]$, which is expected for a chain obeying Gaussian statistics.

It is interesting to note that the three unknowns in Eq. (39) can be determined by three measurements. Since, in general, $K_{neutron} \neq K_{light} \neq K_{x-rays}$, Fedorov and Serdyuk[44] pointed out that if the Q ranges can be arranged to overlap, a measurement on one sample by the three techniques is sufficient to determine $R_g(1)$, $R_g(2)$, and G_{12}.

There has recently been some interest in triblock copolymers, in particular, a PSH–PSD–PSH copolymer where the PSD sequence was only 12% by weight of the whole molecule. A solution of such chains shows a peak in the scattering function which has been associated by de Gennes, using random phase approximations, with a "correlation hole" around the labeled center. The idea of a correlation hole is discussed further in the following section on polyelectrolytes. Duplessix $et\ al.$[46] examined a series of solutions containing a fixed total concentration of hydrogenous and partially labeled chains but varying the relative amounts. Figure 16 shows the extrapolation to zero concentration of labeled chains to obtain the single-particle scattering function $P_1(Q)$ for the labeled section. The corresponding Zimm plot gives a radius of gyration much larger than that expected for a single chain of the same length as the labeled section.

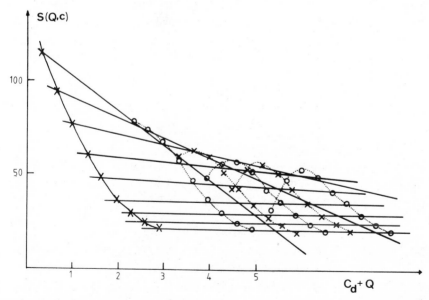

FIGURE 16. Extrapolation to zero concentration of varying concentrations of partially labeled chains in a fixed total solution concentration for polystyrene in CS_2. (From Reference 82.)

3.2.4. Polyelectrolytes in Aqueous Solution

Polyelectrolytes are polymers having one or more ionizable group per monomer unit, e.g., —COOH, which in aqueous solution, dissociate to give a multivalent ionized polymer chain (polyion) plus a large number of small counterions. The properties of polyelectrolytes in solution are influenced by a number of factors, of which the main ones are: the degree of neutralization of the ionizable groups α_s, the polyelectrolyte concentration, the charge density, and the ionic strength of the solution.

It has been known for some time that the end-to-end distance of a polyelectrolyte in aqueous solution decreases as the concentration or the ionic strength increases.[47] In order to account for this behavior and other properties such as the non-Newtonian viscosity it is necessary to assume a certain amount of rigidity in the chain. Moan and Wolff[48] used SANS to investigate this idea and measured the persistence length of carboxymethylcellulose (CMC) in D_2O as a function of α_s and concentration. The CMC was neutralized with NaOH to form Na–CMC and α_s was varied from 0.2 to 1. The CMC used was a high-molecular-weight sample ($M_v = 300,000$) in the concentration range 2.57×10^{-2} to 10^{-3} g/cm³, therefore at the higher concentrations there was probably some overlap of the chains. The effect of neutralization on the intermediate scattering is shown in Figure 17. As the degree of neutralization increases the break in the scattering curve moves to lower Q. This is also found to be the case with decreasing concentration. The data were measured in the intermediate Q range and show the cross-over from Debye behavior where IQ^2 is constant to rigid rod behavior where $IQ^2 \propto Q$. The cross-over point Q^* is a measure of the persistence length l' as

$$l' = 1/Q^*$$

Therefore the persistence length increases with dilution and charge density. Iso-ionic dilution, whereby the charge density is kept constant with dilution by adding a simple electrolyte, e.g., NaCl, has no effect on the persistence length.

The concentration dependence of the persistence length of CMC indicated that at very low concentrations, the chain may well be fully extended and rodlike. However, at the very low concentrations which would be necessary the signal for CMC is not large enough; therefore Moan and Wolff[49] chose a low-molecular-weight poly(methacrylic acid) (PMA) ($M_v = 13,000$) neutralized with NaOH to produce the sodium salt (Na–PMA). Using PMA it is possible to measure a reasonable signal down to concen-

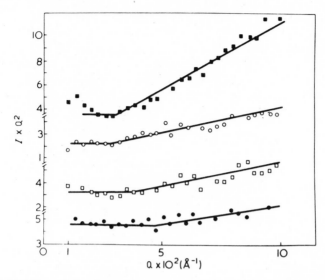

FIGURE 17. Effect of the charge density on the variation of IQ^2 with Q. ■, $\alpha_s = 1$; ○, $\alpha_s = 0.6$; □, $\alpha_s = 0.4$; ●, $\alpha_s = 0.2$. [Reprinted with permission from M. Moan and C. Wolff, *Polymer* **16**, 776 (1975).]

trations of 5×10^{-4} g/cm^3. Measurements were made of R_g as a function of c and α_s. R_g was found to vary linearly with $c^{-1/2}$ for concentrations less than 10^{-2} g/cm^3. The $c^{-1/2}$ dependence is characteristic of a polyion expanding due to the electrostatic repulsions which increase with dilution. From the intermediate scattering it was possible to observe the conformation of the chain changing from Gaussian to rigid rodlike behavior as c decreased. The measurements of R_g as $f(c)$ were made with $\alpha_s = 0.27$. If α_s is increased the chains start to elongate, eventually becoming rodlike, and above a certain concentration c_0, a peak appears in the scattering as shown in Figure 18.[50] c_0 is the concentration below which a rodlike molecule would be able to rotate freely.

The peak shown in Figure 18 has been observed in the scattering from other polyelectrolyte solutions, including the sodium salts of poly(α-L-glutamic acid) (NaGU)[51] in D_2O and deuterated poly(styrene sulfonate)[52] (Na–PSSD) in ultrapure water. Nierlich *et al.*[52] found that the purity of the water was extremely important in determining the shape of the scattering curve as $Q \to 0$ and this is illustrated in Figure 19. Nierlich *et al.*[52] observed that for Na–PSSD in very pure water above c_0 then the intensity as $Q \to 0$ tends to a small value. The small value of $S(Q \to 0)$ is evidence of the osmotic incompressibility of polyelectrolyte solutions. However, the

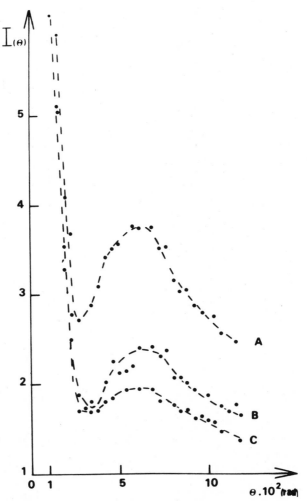

FIGURE 18. Variation of intensity $I(\theta)$ as a function of scattering angle (θ): $\alpha_s = 0.6$, curve A; $\alpha_s = 0.9$, curve B; and $\alpha_s = 1$, curve C. c is 10^{-2} g cm^{-3} ($c_0 \leq 2.5 \times 10^{-3}$ g cm^{-3}). [Reprinted with permission from J. P. Cotton and M. Moan, *J. Phys. Lett.* (*Orsay, France*) **37**, L75 (1976).]

scattering at low Q from sodium poly(styrene sulfonate) (Na–PSS) in D$_2$O, in which cation impurities are present, shows a large increase in intensity.

If salt (NaBr) is deliberately added to the Na–PSSD/H$_2$O solutions, then the low-Q intensity increases until the peak disappears and a smooth curve results. The position of the peak maximum Q_m does not change with the addition of salt. However, variation of the polyelectrolyte con-

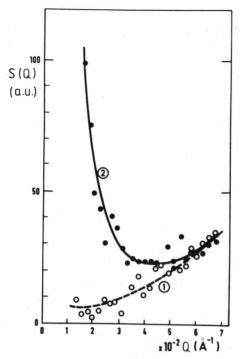

FIGURE 19. Scattered intensity for (1) a solution of deuterated sodium poly(styrene sulfonate) in ultrapure H_2O. $c = 1.96 \times 10^{-2}$ g cm^{-3}; (2) a solution of hydrogenated sodium poly(styrene sulfonate) in D_2O. $c = 1.96 \times 10^{-2}$ g cm^{-3}. The cation impurities measured by mass spectrometry for solution 2 are of order 10^{-4} g cm^{-3} so that $S(Q \to 0)$ does not tend to a small value. [Reprinted with permission from M. Nierlich, C. E. Williams, F. Boué, J. P. Cotton, M. Daoud, B. Farnoux, G. Jannink, C. Picot, M. Moan, C. Wolff, M. Rinaudo, and P. G. de Gennes, *J. Phys.* (*Orsay, France*) **40**, 701 (1979).]

centration does alter Q_m. Nierlich *et al.*[52] found that Q_m was proportional to $c^{1/2}$. Figure 20 shows the scattering from Na–PSSD in H_2O as a function of polyelectrolyte concentration. The exponent of 1/2 agrees with the predictions of two theoretical models, a lattice model of aligned rods[53] and an isotropic model where the chain is considered to consist of a Gaussian distribution of rigid rods.[54]

The origin of the peak in the scattering was investigated by Williams *et al.*[55] One of the problems with the system Na–PSSD/H_2O is that the signal-to-noise ratio is low so that in dilute solution the signal is too small to make accurate measurement of R_g possible. In earlier measurements on macromolecules in solution and in the bulk, the assumption has always been made that in order to measure the single-particle scattering it is neces-

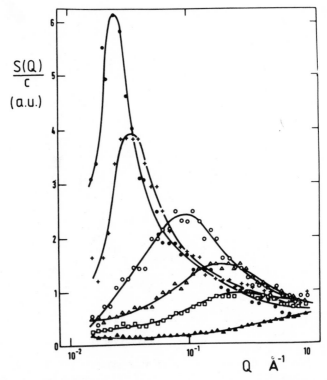

FIGURE 20. Scattered intensity per monomer of a solution of deuterated sodium poly-(styrene sulfonate) ($M_w = 72,000$) in H_2O for solutions of different concentrations. ●, $c = 10^{-2}$ g cm^{-3}; +, 1.96×10^{-2} g cm^{-3}; ○, 4.76×10^{-2} g cm^{-3}; △, 9.09×10^{-2} g cm^{-3}; □, 13.04×10^{-2} g cm^{-3}; ▲, 23×10^{-2} g cm^{-3}. [Reprinted with permission from M. Nierlich, C. E. Williams, F. Boué, J. P. Cotton, M. Daoud, B. Farnoux, G. Jannink, C. Picot, M. Moan, C. Wolff, M. Rinaudo, and P. G. de Gennes, *J. Phys.* (*Orsay, France*) **40**, 701 (1979).]

sary to work at low concentrations where the labeled chains do not overlap. Recently a number of authors have published details of a theoretical treatment which enables the single-particle or intrachain scattering and the interchain scattering to be measured using high concentrations of labeled chains.[15,55,56]

The general equation for the intensity I_c of coherent elastic neutron scattering from a solution of homogenously labeled macromolecules of total number N with Z monomers per chain is obtained by combining equations (15) and (28a) to give

$$I_c = DNZ^2[K_D^{*2}P_1(Q) + NK_D^{*2}P_2(Q)] \tag{40}$$

Consider a solution of a mixture of hydrogenous (H) and deuterated (D) chains which are identical in all respects except their scattering lengths b_H^* and b_D^*, respectively. The total number of chains is N of which XN are deuterated and $(1 - X)N$ are hydrogenous. The intensity from this mixture is given by

$$I_X = D[K_D^{*2}S_{DD}(Q) + K_H^{*2}S_{HH}(Q) + 2K_H^*K_D^*S_{HD}(Q)] \quad (41)$$

Since the chains are identical

$$P_{1,DD}(Q) = P_{1,HH}(Q) = P_1(Q)$$

and

$$P_{2,DD}(Q) = P_{2,HH}(Q) = P_{2,HD}(Q) = P_2(Q)$$

therefore Eq. (41) becomes

$$I_X = DNZ^2\{[K_D^{*2}X + K_H^{*2}(1 - X)]P_1(Q) \\ + N[K_D^*X + K_H^*(1 - X)]^2P_2(Q)\} \quad (42)$$

If the hydrogenous polymer is contrast matched by the solvent then $b_H^* = b_s^*$ so $K_H^* = 0$, therefore

$$I_X = DNZ^2[K_D^{*2}XP_1(Q) + NK_D^{*2}X^2P_2(Q)] \quad (43)$$

Using Eqs. (40) and (43) it is possible to obtain both $P_1(Q)$ and $P_2(Q)$ without using low concentrations

$$I_X/X = DNZ^2K_D^{*2}(1 - X)P_1(Q) + XI_c \quad (44a)$$

and

$$NP_2(Q) = [I_c - I_X/X]/DNZ^2K_D^{*2}(1 - X) \quad (44b)$$

$P_1(Q)$ can be obtained by extrapolating the scattered intensity per monomer to zero content of labeled chains while holding the total concentration constant. $P_2(Q)$ is obtained from the slopes of these extrapolated lines. No assumption has been made about the concentration range so that it is not necessary to work in dilute solution to obtain R_g. To obtain R_g, $P_1(Q)$ must be measured in the range $QR_g < 1$.

Williams et al.[55] used this method to measure P_1 and P_2 for Na–PSS in water. For the solvent they used a mixture of 70% H_2O and 30% D_2O so that there was no contrast between it and the protonated chains. Figure 21 shows the total coherent intensity per monomer unit $I_c(Q, c)/c$ and the

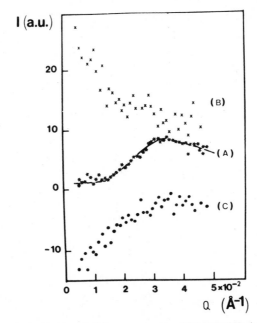

FIGURE 21. Total coherent signal per monomer $I_c(Q, c)/c$ (A), intrachain signal $P_1(Q)$ (B), and interchain signal $P_2(Q)$ (C) for sodium poly(styrene sulfonate) in water, $c = 2 \times 10^{-2}$ g cm^{-3}. [Reprinted with permission from C. E. Williams, M. Nierlich, J. P. Cotton, G. Jannink, F. Boué, M. Daoud, B. Farnoux, C. Picot, P. G. de Gennes, M. Rinaudo, M. Moan, and C. Wolff, *J. Polym. Sci. Polym. Lett. Ed.* **17**, 379 (1979).]

components P_1 and P_2. The single-particle function is a smooth positive curve, whereas the interchain term is negative showing evidence of strong repulsive interactions between the chains. It is these repulsive interactions which give rise to the peak in $I_c(Q)$. While Q_m varies with concentration, little change in R_g was observed other than a possible slight decrease with increasing concentration.

Recent studies of the dynamics of polyelectrolyte solutions around the maximum in $I(Q)$ using neutrons[57] support the idea that Q_m is not a Bragg peak, but is considered to be caused by a correlation hole effect (see, for example, Reference 58). Each polyelectrolyte chain is considered to be surrounded by a correlation tube from which all other chains are strongly expelled. The radius of the tube is the screening length ξ which scales like the interchain distance h'. The correlations between the chains give rise to the following form for $S(Q)$:

$$S(Q) = P_{SR}(Q) \frac{Q^2 h'^2}{(1 + Q^2 h'^2)} \tag{45}$$

where $P_{SR}(Q)$ is the single-rod scattering function. Equation (45) goes through a maximum when $Q = Q_m \approx \xi^{-1}$, which agrees well with the static data of Nierlich et al. The Q-dependent diffusion coefficient $D(Q)$ which was measured using neutron spin-echo decreased sharply below Q_m and was nearly constant above Q_m. An explanation was given in terms of locally rigid chains and a correlation hole effect as described by de Gennes and co-workers.[54,58]

4. DYNAMICS OF MACROMOLECULES IN SOLUTION

4.1. Models and Theoretical Scattering Laws

4.1.1. Simple Dilute Solution

The dynamic scattering laws observed for macromolecules in solution depend on the distance scale explored. For dilute solutions in the range $Q < R_g^{-1}$, overall center-of-mass diffusion is the dominant motion, leading to a very simple expression in the time domain for both coherent and incoherent scattering,

$$S(\mathbf{Q}, t)/S(\mathbf{Q}, 0) = \exp(-\Gamma t) \tag{46}$$

with $\Gamma = DQ^2$, where D is the diffusion coefficient. This exponential time decay becomes a Lorentzian curve when Fourier transformed to the frequency domain (again for both coherent and incoherent scattering):

$$S(\mathbf{Q}, \omega) = \frac{DQ^2}{(DQ^2)^2 + \omega^2} \tag{47}$$

The half-width at half-maximum (HWHM) $\Delta\omega = DQ^2$ and is identical to the inverse correlation time Γ.

Except for the smallest molecules, this motion is inaccessible to neutron scattering experiments. Conventionally, photon correlation spectroscopy has been used to observe diffusion of macromolecules in solution (see Chapter 2 in the work by Pecora). At higher Q, the internal motion of the polymer molecule is explored. The scattering laws based on the Rouse or Zimm models (see Chapter 1, Forsman) of this motion have been derived by Pecora[59] and, in a more explicit form, by de Gennes.[60,61] In the time domain they show an exponential dependence on $t^{1/2}$ and $t^{2/3}$, respectively. Under θ conditions, the inverse correlation time Γ has a simple form in the range $R_g^{-1} \ll Q \ll a^{-1}$, for coherent scattering:

$$\Gamma = \frac{1}{12\sqrt{2}} \frac{k_B T}{G_0} a Q^4 \qquad \text{(Rouse)} \qquad (48a)$$

$$\Gamma = \frac{1}{6\pi\sqrt{2}} \frac{k_B T}{\eta_s} Q^3 \qquad \text{(Zimm)} \qquad (48b)$$

where a is the length associated with one Rouse spring unit and G_0 the friction coefficient of such a unit. η_s is the solvent viscosity, k_B is Boltzmann's constant, and T is the temperature. When Fourier transformed to the frequency domain, the scattering laws no longer have a simple analytical form but $\Delta\omega$ follows the same Q variation with different numerical coefficients

$$\Delta\omega_{\text{coh}} = 0.34\Gamma_{\text{Rouse}} \qquad (49a)$$

$$\Delta\omega_{\text{coh}} = 0.85\Gamma_{\text{Zimm}} \qquad (49b)$$

$$\Delta\omega_{\text{inc}} = 1.1\Gamma_{\text{Zimm}} \qquad (49c)$$

$\Delta\omega_{\text{inc}}$ was not explicitly calculated for the Rouse case. The scattering laws themselves have rather more intensity in the wings than a Lorentzian of the same half-width.[62,13]

For $Q > a^{-1}$ local Brownian motion of short segments dominates and a return to a Q^2 dependence of $\Delta\omega$ is anticipated. The length, a, of a Rouse spring unit has to be large enough for the chain segment connecting two points this distance apart to obey Gaussian statistics. It is therefore expected that, even for the most flexible polymers, a will be greater than 10 Å. Neutron scattering experiments necessarily explore the region $Q \simeq a^{-1}$. This has prompted a number of recent calculations explicitly exploring the scattering from motion in this range. Akcasu et al.[63] developed the theory of scattering from a bead spring model of a polymer in solution to cover continuously the whole range from molecular diffusion through long-range segmental motion (Rouse–Zimm region) to local bond motion (which is represented by the Brownian motion of a single bead). The time domain scattering law is expressed in terms of an inverse correlation time Ω which is, in fact, the first cumulant of $S(\mathbf{Q}, t)$

$$\Omega = -\lim_{t \to 0} \frac{d \ln[S(\mathbf{Q}, t)]}{dt} \qquad (50)$$

In the range $R_g^{-1} < Q < a^{-1}$ the same scattering laws are obtained as for the de Gennes[60,61] calculation and the same dependence of the correlation time with $\Omega \equiv \sqrt{2}\,\Gamma$ time. As the value of Qa increases above unity the correlation functions progressively deviate from the de Gennes curves towards a simple exponential decay at high Q, as shown in Figure 22. All the calculations mentioned so far apply in θ conditions. Akcasu *et al.*[63] show that in a good solvent Ω has the same Q^3 behavior for the Zimm model, but the numerical coefficient increases from $1/6\pi$ to 0.72. This temperature dependence will be discussed further in Section 4.1.2. All the calculations assume a preaveraged Oseen tensor (cf. Chapter 1, Forsman), but it has been shown[64] that nonpreaveraging preserves the Q dependence while changing the numerical prefactor by up to 10%. It is now clear why the time domain results from the neutron spin-echo technique are particularly easy to interpret in terms of these calculations. The initial slope of the scattering function is relatively easily obtained, but while Ω does translate to the frequency domain in terms of $\Delta\omega$, (with an appropriate numerical coefficient) the data have to be analyzed, as described in 2.2.4, by convoluting a complete model scattering law with the resolution func-

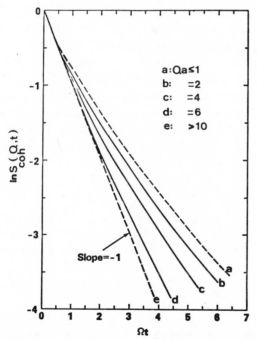

FIGURE 22. Theoretical correlation functions calculated for different values of Qa. $Qa < 1$ corresponds to Zimm motion, $Qa > 10$ corresponds to a simple exponential characteristic of Fickian diffusion. The intermediate values are from Reference 63. (From Reference 70.)

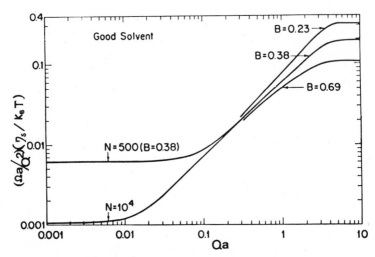

FIGURE 23. Variation of normalized Ω values with normalized Q for different values of the draining parameter B and for two chain lengths in a good solvent. (From Reference 70.)

tion. Such a scattering law is not at present available in the frequency domain for the complete Q range, and data have to be interpreted using either the Lorentzian form or the de Gennes Rouse–Zimm functions which are not applicable in the intermediate Q ranges.

Figure 23 (taken from Reference 63) shows the normalized inverse correlation time $(\Omega\eta_s a)/(Q^2 k_B T)$ plotted against Qa calculated using the Zimm model for different molecular sizes (designated by the number N of bead units), and for different values of the parameter B, where

$$B = \frac{1}{(6\pi^2)^{1/2}} \frac{G_0}{\eta_s a} \tag{51}$$

or $B = \sqrt{2}h/\sqrt{N}$, where h is the draining parameter introduced by Zimm (discussed by Forsman in Chapter 1). The three regions of dilute solution behavior are clearly seen as the three sections of the S-shaped curve.

4.1.2. Temperature and Concentration Effects

Paralleling the theoretical developments for static scaling laws, discussed in Section 3.2.2a, there have been a number of predictions of the effect of concentration and temperature on dynamic properties. Although the experimental data in Section 4.3 are, as yet, very limited, it is worth outlining the predictions at this point for this, a fast developing area.

For the range $QR_g > 1 > Qa$, Akcasu and Benmouna[65] demonstrated that there would be cross over in the Q dependence of Ω, for a polymer in dilute solution, at a temperature dependent Q^*,

$$Q^* = \sqrt{6}/\xi_\tau \qquad (52)$$

where ξ_τ is a temperature-dependent screening length, and is a measure of the distance beyond which excluded-volume interaction between monomers becomes important. Its temperature dependence is just given by $T/(T - \theta)$. In the Zimm limit, as has been mentioned, Ω shows the same Q^3 dependence above and below Q^*. For $Q < Q^*$, excluded-volume effects are important and

$$\Omega = 0.071 \frac{k_B T}{\eta_s} Q^3 \qquad (53)$$

Above Q^* the behavior of Q is the same as that for θ conditions and the numerical prefactor drops to 0.053 [Eq. (48b) with $\Omega = \sqrt{2}\Gamma$]. The curves shown in Figure 23 would thus show a step of $0.071/0.053$ at some value of Q in the sloping region for intermediate solvents. Akcasu and Benmouna[66] also discuss some of the effects of concentration. If the single-chain dynamics are observed (by labeling) then in the Zimm limit Ω again displays a transition, but this time in the reverse direction, from $0.053 Q^3 k_B T/\eta_s$ for $Q < Q^*$ to $0.071 Q^3 k_B T/\eta_s$ for $Q > Q^*$. The concentration-dependent $Q^* = 6^{1/2}/\xi$, where ξ is the screening length discussed in Section 3.2.2a. For scattering from identical chains, the authors predict the concentration dependence of the diffusion coefficient at low Q. These results are more relevant to light scattering experiments and there are no predictions for the higher-Q regime. de Gennes,[67] however, showed that for $Q < Q^*$, in the regime $QR_g > 1 > Qa$, viscous modes will exist so that the system is reminiscent of a gel with a cooperative diffusion coefficient $D_c = k_B T/(6\pi\eta_s\xi)$.

The inverse correlation time for such a gel will show a Q^2 dependence, but for $Q > Q^*$ a return to Q^3 behavior is predicted.

4.2. Results

4.2.1. Small-Chain Dynamics—Noninteracting Systems

Earlier neutron scattering experiments using time-of-flight or back-scattering techniques attempted to obtain evidence for Rouse or Zimm motions of polymer chains in solution in the frequency domain. It became clear, however, that even with the high resolution, at low Q, of the back-scattering spectrometer, it was impossible to obey the upper inequality $Q \ll a^{-1}$, so that some sort of intermediate behavior was obtained with $\Delta\omega$ varying as Q^n, where $2 < n < 3$.[13,68] These frequency domain data were analyzed, as described in Section 2.2.4, by convoluting the model $S(\mathbf{Q}, \omega)$ with the instrumental resolution function and using $\Delta\omega$ as a variable parameter. Despite the difficulties inherent in fitting data in this way, there was strong evidence that scattering laws based on the Zimm model (i.e., including hydrodynamic interactions) gave the best fit to the experimental points.[13,68,69]

The increased resolution and smaller Q values accessible with the neutron spin-echo spectrometer have considerably improved this situation. Analysis of the observed correlation functions shows the nonlinear time dependence of $\ln S(\mathbf{Q}, t)$ expected for the intermediate Q range.[70,71] For most polymers, however, Ω does not show a pure Q^3 dependence indicating that Q values are still not low enough to obey $Q \ll a^{-1}$. Poly(dimethyl siloxane) in C_6D_6 is apparently flexible enough and the consequent length parameter, a, short enough to show Q^3 dependence for Q values in the range 0.02–0.1 Å^{-1}. Polytetrahydrofuran and polystyrene in both C_6D_6 and CS_2 both show clearly behavior characteristic of the transition region $\mathbf{Q} \approx a^{-1}$. Although Richter et al. forced-fit Q^3 behavior to data from poly-(methyl methacrylate) in C_6D_6, they remark on the nonuniversal nature of the proportionality constant between Ω and $(k_B T/\eta_s)Q^3$. This is almost certainly a consequence of assuming the Q^3 power law where it does not apply. Assuming the de Gennes correlation function at higher Qa values will have the effect of overstimating Ω as can be seen from Figure 22.

If the draining parameter B is fixed, at a value found reasonable for fitting high-frequency viscoelastic data,[72] then the variable a can be obtained by a sliding fit of data along the two axes in Figure 23, all other parameters being fixed.[70] Figure 24 shows data for three polymers [polystyrene, poly-tetrahydrofuran, and poly(dimethyl siloxane)] in C_6D_6 at 30 °C, each normalized by a fitted value of a to lie on the calculated curve. The full line is calculated for good solvent conditions (which nominally apply for all three polymers); the dotted curve is the θ solvent line. It is interesting to note that at values of $Qa > 3$, where much of the neutron data lie, the solvent quality has no effect. However, there is a noticeable tendency for the PDMS data, for which $Qa < 1$, to fall closer to the θ solvent line although the experiment was carried out at 40 °C above the theta temperature. The effects of temperature will be further discussed in Section 4.2.2.

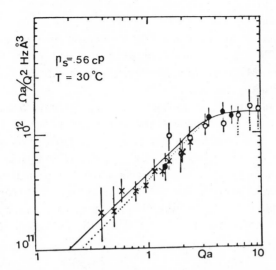

FIGURE 24. ln–ln plot of $\Omega a/Q^2$ against Qa for three polymers in C_6D_6. $\bigcirc$, polystyrene; $\bullet$, polytetrahydrofuran; $\times$, poly(dimethyl siloxane). The solid curve is the theoretical curve for a good solvent of viscosity 0.56 cP at 30 °C and with $B = 0.38$ (Reference 63). The dotted curve is the corresponding θ solvent curve. (From Reference 70.)

The relatively short distance scale characteristic of the neutron experiments has recently been explored from the point of view of a rather different model. Allegra and Canazzoli[73] showed that it is possible to select a range of frequencies where the anharmonicity of the intramolecular potential of a "real" polyethylene chain translates into a model where first, second, third, etc. nearest neighbours in the skeletal backbone are connected by progressively weaker harmonic springs. They calculated numerically the inverse correlation time Γ and the first cumulant Ω as a function of Q in the limited range $0.1 < Q < 0.4$ Å^{-1}, without hydrodynamic interactions, and showed a variation Q^n where $n \simeq 3.6$ for coherent scattering. Although this apparent power law falls below the asymptotic value $n = 4$ for Rouse chains without hydrodynamic effects, it is much higher than the experimentally determined Q variation for this range (as in Figure 24), probably indicating the importance of hydrodynamic effects for macromolecules in solution even over these short distances. The values of a determined from the data in Figure 24 are 55, 45, and 16 Å for PS, PTHF, and PDMS, respectively.

These lengths should not be interpreted too literally in terms of the length of a Rouse spring, but they do give an idea of the length scale associated with the onset of these Rouse modes and its variation with polymer structure. There is quite a good agreement between the neutron values, and the values of the spring length obtained from viscoelastic measurements[70,72] (other results from techniques which inherently explore local motion—NMR relaxation, fluorescence depolarization—compare very reasonably with the neutron data[70]).

It is notable that for a polymer such as polystyrene a is rather long. Examination of Figure 23 shows that significant deviations from the Q^3 asymptote are observed for values of $Qa > 2$. On the other hand, the lower limit to the Q^3 region is set by $QR_g \simeq 1$. For smaller polystyrene molecules with $R_g < 50$ Å or for higher molecular weight but very stiff polymers, where $a \gg 50$ Å, the Q^3 region may well disappear altogether.

An example of the lower limit $QR_g \sim 1$ appearing within the neutron range can be seen in Figure 25. This shows the analogous Ω/Q^2 plot to

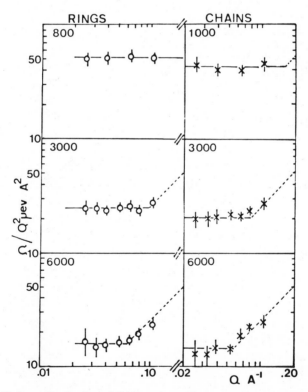

FIGURE 25. Logarithmic plots of Ω/Q^2 against Q for a sequence of low-molecular-weight poly(dimethyl siloxane) molecules of ring and chain form in 3% solution in C_6D_6 at 30 °C. R_g values in Å for (i) rings: 800, < 5; 3000, 9.54; 6000, 14.3; and (ii) chains: 1000, < 6; 3000, 13.1; 6000, 19.9.

TABLE 3. Values of R_g, R_H, and R^* for Cyclic (R) and Linear (L)
Poly(dimethyl siloxane) Chains

Sample	800 R	300 R	6000 R	1000 L	3000 L	6000 L
M_n	740	7401	6290	888	2738	6364
R_g (Å)		9.5	14.3		13.1	19.9
R_H (Å)	5	10.5	16.3	6.0	12.1	17.6
R^* (Å)	5	10.0	18.0	6.3	12.5	20

Figure 24 for low-molecular-weight poly(dimethyl siloxane) molecules in cyclic and linear forms in dilute solution in C_6D_6.[74,75] From the horizontal region the diffusion coefficient D is obtained directly and hence, assuming the Stokes–Einstein relationship, $D = k_B T/(6\pi\eta_s R_H)$, the hydrodynamic radius R_H can be calculated. The turn round point, denoted by R^*, from the horizontal to the sloping line, should be closely related to the molecular dimensions. Table 3 lists R_H, R^*, and R_g.

The ratio of R_g(ring)/R_g(chain) has been discussed in the previous section; it is around 0.707. The corresponding ratio for R_H is calculated[63] to be 0.865. The ratios observed show a reasonable agreement with this value and, as expected, R^* follows the pattern of R_g values.

Figure 23 thus maps out behavior for the dynamics of a macromolecule in dilute solution which is indeed observed in practice, and by judicious choice of the polymer structure and dimensions, all the different ranges of behavior can be observed in the somewhat limited experimental Q range. The importance of the neutron results is their simultaneous provision of frequency and distance scale information.

4.2.2. Effects of Polymer–Solvent and Polymer–Polymer Interaction

In the good solvent regime where the quality of the solvent can improve no further, increases in temperature should simply increase Ω by the factor T/η_s in Eq. (48b). Nicholson et al.[70] showed that this is approximately the case for polystyrene in C_6D_6. Since most of the neutron data fall in the range $Qa \gtrsim 1$, rather more complex effects of temperature might be expected even if the solvent quality remains constant. Changes in chain flexibility might alter the length parameter a and changes in the draining parameter B cannot be ruled out.

When the solvent quality is also changing, between its θ and good solvent limits, further complications may ensue. Poly(dimethyl siloxane) is the only polymer for which a reasonable amount of data lies below $Qa = 1$. It has already been remarked that the θ data at 30°C lie close to the θ-solvent values despite being about 40°C above the θ temperature of C_6D_6. Similar effects have been observed at lower Q values for polystyrene in various solvents from light scattering experiments.

Akcasu and Han showed that the ratio of the hydrodynamic radius R_H (obtained from measurements of diffusion coefficients) to the static radius R_g does not reach its full good solvent value until temperatures considerably (> 70°C) above the theta temperature. The dynamic behavior of polymer chains is slow to show the effects of excluded volume as the temperature is increased from θ. Working at somewhat higher Q values and with the difficulties in analyses inherent in the frequency domain data from the back-scattering spectrometer (Section 2.2.4), Ewen et al.[77] explored the temperature dependence of Ω for PDMS in benzene and chlorobenzene between 260°K and 350°K. For all temperatures close agreement with Eq. (48b) was found, and the small deviations are certainly accounted for by the fact that the data extend to $Q = 0.3$ Å^{-1}, where $Qa > 1$.

For polystyrene in cyclohexane, however, spin-echo data do show a change in the numerical prefactor in Eq. (48b) as temperature increases from the theta conditions.[78] There are, though, a number of discrepancies between these data and theoretical predictions. Notably, the transition is apparently extremely sharp at low Q and is very quickly smeared out and disappears at higher Q values while the magnitude of the effect is too large by a factor of 3 or more. The identity of the θ and good solvent predictions for Ω when $Qa > 3$ is seen clearly in Figure 24 so that the smearing out of the cross-over at higher Q is expected. Moreover, polystyrene has been shown (see Section 4.2.1) to have a large a value due to the relative chain stiffness, so that Qa is probably greater than unity for the entire Q range explored. Forcing Q^3 behavior to the data will undoubtedly produce odd effects in the numerical prefactors, and may accentuate the apparent sharpness of the change at low Q. The values of Q^* are rather smaller than the values extracted from static measurements on the same system but, for the reasons described above, this difference may not be important.

In this Q range changes in a and B would strongly affect Ω and cannot be ruled out. It is worth noting that there is considerable evidence for a flexibility change in polystyrene at around 60°C[79,80] (see discussion of Pethrick in this volume) and such a change might well be expected to show up in the effective spring length, a.

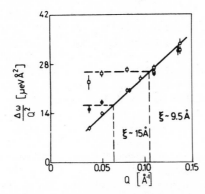

FIGURE 26. Width function $\Delta\omega/Q^2$ against Q for three concentrations of linear poly(dimethyl siloxane) in benzene-d_6. $\bigcirc$, 0.05 g/cm³; $\bullet$, 0.15 g/cm³; $\square$, 0.30 g/cm³. [Reprinted with permission from D. Richter, B. Ewen and J. B. Hayter, *Phys. Rev. Lett.* **41**, 1484 (1978).]

The extent of the Q^3 range observable for poly(dimethyl siloxane) makes it a good candidate for observation of low Q deviations from this behavior. The effect of center-of-mass diffusion for small molecules was shown in Figure 25. As the concentration increases, for high M_w samples, gel-like Q^2 behavior of Ω is expected for distances $Q^{-1} > \xi$ (the screening length). Richter et al.[71] observed such behavior for PDMS of molecular weight 30,000 in C_6D_6 as the concentration increased to 30%. Figure 26 shows a change from Q^3 to Q^2 behavior as expected. The values of ξ extracted, assuming $\xi Q^* = 1$, are in the same range as values from static experiments[37] on polystyrene in CS_2 and, with a fairly large experimental error, ξ varies as $c^{3/4}$.

It is worth remarking here that when there are very strong interactions within the system—for example, in charged systems—producing peaks in the structure factor $S(Q)$ these are reflected in the dynamic behavior. An example of this is the scattering from polyelectrolyte solutions discussed in Section 3.2.4. Another example (beyond the scope of this chapter) where neutron spin-echo results are providing definitive structural information is that of micelle solutions.[81]

REFERENCES

1. B. T. M. Willis (ed.), *Chemical Applications of Thermal Neutron Scattering*, Oxford University Press, London (1973).
2. G. Kostorz (ed.), *Treatise on Materials Science and Technology*, Vol. 15, Neutron Scattering, Academic Press, New York (1979).

2a. J. S. Higgins, in *Treatise on Materials Science and Technology* (G. Kostorz, ed.), Vol. 15, pp. 381–422, Academic Press, New York (1979).

3. J. S. Higgins and R. S. Stein, *J. Appl. Crystallogr.* **11**, 346–375 (1978).

4. A. Maconnachie and R. W. Richards, *Polymer* **19**, 739–762 (1978).

5. B. Jacrot, *Rep. Prog. Phys.* **39**, 911–953 (1976).

6. H. Stuhrmann and A. Miller, *J. Appl. Crystallogr.* **11**, 325–345 (1978).

7. G. G. Kneale, J. P. Baldwin, and E. M. Bradbury, *Q. Rev. Biophys.* **10**, 485–527, (1977).

8. W. Schmatz, T. Springer, J. Schelten, and K. Ibel, *J. Appl. Crystallogr.* **17**, 96–116 (1974).

9. M. Birr, A. Heidemann, and B. Alefeld, *Nucl. Instrum. Methods* **95**, 435–439 (1971).

10. F. Mezei, *Z. Phys.* **255**, 146–160 (1972).

11. J. B. Hayter, in *Neutron Diffraction* (H. Dachs, ed.), Springer-Verlag, Berlin (1978).

12a. P. Dagleish, J. B. Hayter, and F. Mezei, in *Neutron Spin-Echo*, (F. Mezei, ed.), Physics 128, Springer-Verlag, Berlin (1980).

12b. J. B. Hayter in *Neutron Spin-Echo* (F. Mezei, ed.), Physics 128, Springer-Verlag, Berlin (1980).

13. J. S. Higgins, G. Allen, R. E. Ghosh, W. S. Howells, and B. Farnoux, *Chem. Phys. Lett.* **49**, 197–202 (1977).

14. G. C. Summerfield, in *Spectroscopy in Biology and Chemistry, Neutron, X-ray and Laser* (S. H. Chen and S. Yip, eds.), Chap. 10, Academic Press, New York (1974).

15. H. Benoit, J. Koberstein, and L. Leibler, *Makromol. Chem. Suppl.* **4**, 85 (1981).

16. B. H. Zimm, *J. Chem. Phys.* **16**, 1093–1099 (1948).

17. P. Debye, *J. Phys. Colloid Chem.* **51**, 18–32 (1947).

18. A. Guinier and G. Fournet, *Small Angle Scattering of X-Rays*, Wiley, New York (1955).

19. D. G. H. Ballard, M. G. Rayner, and J. Schelten, *Polymer* **17**, 349–351 (1976).

20. S. Heine, O. Kratky, and G. Porod, *Makromol. Chem.* **44**, 682–726 (1961).

21. J. S. Higgins, K. Dodgson, and J. A. Semlyen, *Polymer* **20**, 553–558 (1979).

22. P. Kratochvil, in *Light Scattering from Polymer Solutions* (M. B. Huglin, ed.), Academic Press, New York (1972).

23. A. K. Gupta, J. P. Cotton, E. Marchal, W. Burchard, and H. Benoit, *Polymer* **17**, 363–366 (1976).

24. S. F. Edwards, *Proc. Phys. Soc. London* **88**, 265–280 (1966).

25. P. J. Flory, *Principles of Polymer Chemistry*, Cornell University Press, Ithaca, New York (1953).

26. P. G. de Gennes, *Phys. Lett. A* **38**, 339–340 (1972).

27. J. des Cloiseaux, *J. Phys. (Orsay, France)* **36**, 281–291 (1975).

28. P. G. de Gennes, *J. Phys. Lett. (Orsay, France)* **46**, L55–L57 (1975).

29a. M. Daoud and G. Jannink, *J. Phys. (Orsay, France)* **37**, 973–979 (1976).

29b. M. Daoud, *J. Polym. Sci. Polym. Symp.* **61**, 305–311 (1977).

30. S. F. Edwards, *J. Phys. A.* **8**, 1670–1680 (1975).

31. S. F. Edwards and E. F. Jeffers, *J. Chem. Soc., Faraday Trans. 2* **75**, 1020–1029 (1979).

32. C. Strazielle and H. Benoit, *Macromolecules* **8**, 203–205 (1975).

33. J. P. Cotton, M. Nierlich, F. Boué, M. Daoud, B. Farnoux, G. Jannink, R. Duplessix, and C. Picot, *J. Chem. Phys.* **65**, 1101–1108 (1976).

34a. R. W. Richards, A. Maconnachie, and G. Allen, *Polymer* **22**, 147–152 (1981).

34b. R. W. Richards, A. Maconnachie, and G. Allen, *Polymer* **22**, 153–157 (1981).
34c. R. W. Richards, A. Maconnachie, and G. Allen, *Polymer* **22**, 158-162 (1981).
35. B. Farnoux, M. Daoud, D. Decker, G. Jannink, and R. Ober, *J. Phys. Lett.* (*Orsay, France*) **36**, L35–L39 (1975).
36. B. Farnoux, F. Boué, J. P. Cotton, M. Daoud, G. Jannink, M. Nierlich, and P. G. de Gennes, *J. Phys.* (*Orsay, France*) **39**, 77–86 (1978).
37. M. Daoud, J. P. Cotton, B. Farnoux, G. Jannink, G. Sarma, H. Benoit, R. Duplessix, C. Picot, and P. G. de Gennes, *Macromolecules* **8**, 804–818 (1975).
38. R. W. Richards, A. Maconnachie and G. Allen, *Polymer* **19**, 266–270 (1978).
39. D. W. Schaefer, J. F. Joanny, and P. Pincus, *Macromolecules* **13**, 1280–1289 (1980).
40. C. C. Han and B. Mozer, *Macromolecules* **10**, 44–51 (1977).
41. M. Leng and H. Benoit, *J. Polym. Sci.* **57**, 263–273 (1962).
42. M. Duval, R. Duplessix, C. Picot, D. Decker, P. Rempp, H. Benoit, J. P. Cotton, R. Ober, G. Jannink, and B. Farnoux, *J. Polym. Sci. Part B* **14**, 588–589 (1976).
43. L. M. Ionescu, Ph. D. thesis, University of Strasbourg (1976).
44. I. N. Serdyuk and B. A. Dedorov, *J. Polym. Sci. Polym. Lett. Ed.* **11**, 645–649 (1973).
45. P. G. de Gennes, *J. Phys.* (*Orsay, France*) **31**, 235–238 (1970).
46. R. Duplessix, J. P. Cotton, H. Benoit, and C. Picot, *Polymer* **20**, 1181–1182 (1979).
47. C. Tanford, *Physical Chemistry of Macromolecules*, Wiley, New York (1961).
48. M. Moan and C. Wolff, *Polymer* **16**, 776–780 (1975).
49. M. Moan and C. Wolff, *Polymer* **16**, 781–784 (1975).
50. J. P. Cotton and M. Moan, *J. Phys. Lett.* (*Orsay, France*) **37**, L75–L77 (1976).
51. M. Rinaudo and A. Domard, *J. Polym. Sci. Polym. Lett. Ed.* **15**, 411–415 (1977).
52. M. Nierlich, C. E. Williams, F. Boué, J. P. Cotton, M. Daoud, B. Farnoux, G. Jannink, C. Picot, M. Moan, C. Wolff, M. Rinaudo, and P. G. de Gennes, *J. Phys.* (*Orsay, France*) **40**, 701–704 (1979).
53. S. Lifson and A. Katchalsky, *J. Polym. Sci.* **13**, 43–55 (1954).
54. P. G. de Gennes, P. Pincus, R. M. Velasco, and F. Brochard, *J. Phys.* (*Orsay, France*) **37**, 1461-1473 (1976).
55. C. E. Williams, N. Nierlich, J. P. Cotton, G. Jannink, F. Boué, M. Daoud, B. Farnoux, C. Picot, P. G. de Gennes, M. Rinaudo, M. Moan, and C. Wolff, *J. Polym. Sci., Polym. Lett. Ed.* **17**, 379–384 (1979).
56. A. Z. Akcasu, G. C. Summerfield, S. N. Jahshan, C. C. Han, C. Y. Kim, and H. Yu, *J. Polym. Sci. Polym. Phys. Ed.* **18**, 863–869 (1980).
57. J. Hayter, G. Jannink, F. Brochard-Wyart, and P. G. de Gennes, *J. Phys. Lett.* (*Orsay, France*) **41**, L451–L454 (1980).
58. P. G. de Gennes, *Scaling Concepts in Polymer Physics*, Cornell University Press, Ithaca, New York (1979).
59. R. Pecora, *J. Chem. Phys.* **49**, 1032–1035 (1968).
60. P. G. de Gennes, *Physics* **3**, 37–45 (1967).
61. E. du Bois Violette and P. G. de Gennes, *Physics* (*N. Y.*) **3**, 181–198 (1967).
62. J. S. Higgins, R. E. Ghosh, W. S. Howells, and G. Allen, *J. Chem. Soc. Faraday Trans. 2* **73**, 40–47 (1977).
63. A. Z. Akcasu, M. Benmouna, and C. C. Han, *Polymer* **21**, 866–890 (1980).
64. M. Benmouna and A. Z. Akcasu, *Macromolecules* **13**, 409–414 (1980).
65. M. Benmouna and A. Z. Akcasu, *Macromolecules* **11**, 1187–1192 (1978).
66. A. Z. Akcasu and M. Benmouna, *Macromolecules* **11**, 1193–1198 (1978).
67. P. G. de Gennes, *Macromolecules* **9**, 594–598 (1976).

68. G. Allen, R. Ghosh, J. S. Higgins, J. P. Cotton, B. Farnoux, G. Jannink, and G. Weill, *Chem. Phys. Lett.* **38**, 577–581 (1976).
69. A. Z. Akcasu and J. S. Higgins, *J. Polym. Sci. Polym. Phys. Ed.* **15**, 1745–1756 (1977).
70. L. K. Nicholson, J. S. Higgins, and J. B. Hayter, *Macromolecules* **14**, 836–843 (1981).
71. D. Richter, J. B. Hayter, F. Mezei, and B. Ewen, *Phys. Rev. Lett.* **41**, 1484–1487 (1978).
72. K. Osaki and J. L. Schragg, *Polym. J.* **2**, 541–549 (1971).
73. G. Allegra and F. Ganazzoli, *J. Chem. Phys.* **74**, 1310–1320 (1981).
74. J. S. Higgins, L. K. Nicholson, and J. B. Hayter, *Polym. Prepr. Am. Chem. Soc. Div. Polym. Chem.* **22**, 86–88 (1981).
25. J. S. Higgins, K. Dodgson, and J. A. Semlyen, *Polymer* **20**, 553–558 (1979).
76. A. Z. Akcasu and C. C. Han, *Macromolecules* **12**, 276–280 (1979).
77. B. Ewen, D. Richter, and B. Lehnen, *Macromolecules* **13**, 876–880 (1980).
78. D. Richter, B. Ewen, and J. B. Hayter, *Phys. Rev. Lett.* **45**, 2121–2125 (1980).
79. C. Reiss and H. Benoit, *C. R. Hebd. Seances Acad. Sci. Ser. C.* **253**, 268–270 (1961).
80. J. H. Dunbar, A. M. North, R. A. Pethrick, and P. B. Teik, *Polymer* **21**, 764–768 (1980).
81. J. B. Hayter and J. Penfold, *J. Chem. Soc. Faraday Trans. 1* **77**, 1851 (1981).
82. R. Duplessix, International Union of Pure and Applied Chemistry, Macromolecular Symposium, Mainz (1979), abstracts, p. 870.

5

Molecular Motion in Concentrated Polymer Systems: High-Frequency Behavior

RICHARD A. PETHRICK

1. INTRODUCTION

An understanding of the various types of molecular motion possible in polymer solutions has recently become possible by the correlation of various types of experimental observation with the predictions of theory. How changes in concentration influence the spectrum of high-frequency motions in solution and in the melt will be the underlying theme of this chapter.

Studies on dilute solutions are usually performed in an attempt to characterize the way in which intramolecular interactions influence segmental and other high-frequency motions. In dilute solution, strong polymer–polymer interactions are replaced by the weaker polymer–solvent interactions. The connection between the shear viscosity and molecular motion is discussed in Chapter 1. In dilute solution, "normal mode" theories[1,2] can adequately describe the relaxation of an "isolated" polymer molecule in solution; however, increasing the concentration leads to

RICHARD A. PETHRICK ● Department of Pure and Applied Chemistry, University of Strathclyde, Glasgow G1 1XL, Scotland, United Kingdom.

significant departure from ideal behavior and requires the introduction of terms associated with polymer–polymer interactions. The effects of change in concentration can be classified either using the semiempirical rules of Frisch and Simha[3] or the predictions of Edwards[4] based on the concepts of "hydrodynamic screening" or of de Gennes[5] based on scaling concepts. The essential concepts are common to all the approaches mentioned.

In dilute solution, defined as a concentration where the product of the intrinsic viscosity $[\eta]$ multiplied by the concentration c is less than one, the polymer molecule acts as an isolated chain. The effect of the polymer on the solvent flow can be predicted on the basis of a simple additivity and is the result of inelastic collisions between a Brownian thermal continuum and a polymer chain which obeys Gaussian statistics. In high-molecular-weight polymers, interactions between elements within the same chain lead to the so-called "excluded-volume" effects. The theory associated with such interactions has been described in Chapter 1.

Increasing the concentration raises the product $[\eta]c$ to a value of between 1 and 4, and allows polymer–polymer intermolecular interactions to influence the motion of the polymer. These interactions can be invisaged to be of two types; firstly because of the relative close proximity of neighboring chains, the perturbation of the lamellae flow by the presence of the polymer has not died down before the next polymer molecule is encountered, and secondly, the occurrence of intermolecular interactions which have a similar effect to the intramolecular interactions associated with excluded-volume effects observed in dilute solutions. Both these effects can be treated theoretically by the introduction of a "hydrodynamic screening" interaction. The moderately concentrated region would be expected to extend from the point where hydrodynamic flow ceases to be lamellae to a point where the polymer completely fills the solution, but does not lead to significant interpenetration of the volume occupied by a neighboring polymer molecule (Figure 1). According to Frisch and Simha[3] the upper limit of this region corresponds to a condition where $[\eta]c = 4$. Edwards[4] treats this region in terms of the effects of a characteristic screening length (ξ) which describes the hydrodynamic interactions. An extension of the original theory of Zimm[1] has recently been published[6] and allows a smooth interpolation between the dilute and moderately concentrated solution limits, the higher concentration limit corresponds to Rouse theory.[2] The theory[6] introduces a hydrodynamic interaction parameter which varies with concentration and in so doing changes the eigenvalues and vectors used in the theoretical formulation of dilute solution behavior. The parallel is obvious between this approach and that of Edwards.[4]

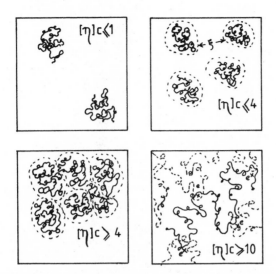

FIGURE 1. Representation of the Frisch and Simha Regions of polymer–polymer interaction. Dotted circles represent the effective volume occupancy of a polymer coil; ξ is the screening length. Dashed lines represent the surrounding polymer matrix; e represents a point of entanglement.

At even higher concentrations, $[\eta]c = 10$, intercoil contact has been exchanged for dynamic contacts as a consequence of coil–coil interpenetration (Figure 1) and a pseudo-matrix-gel is formed. The above limit $[\eta]c \approx 10$ can be considered as being indicative of entanglement formation. Gel formation will occur when the time-averaged number of entanglements per chain exceeds a value of 1.5. The probability chain–chain contacts occurring depends upon the size of the coil and hence on the end-to-end distance for the polymer (Figure 1).

Edwards[4] and de Gennes[5] have computed the regions of behavior for a polymer in solution using mean-field hydrodynamic screening (MF) and scaling law (SL) theories (Figure 2). The axes are chosen to overcome the problem of differences in the thermodynamic quality of the solvent and the differences in molecular weight of the polymer.[7,8] The axes chosen are, respectively, $\phi = (T - \theta)/\theta$, where θ is the Flory temperature and is indicative of ideal interactions between polymer and solvent, and $cM^{1/2}$, where c is the concentration of polymer. Consequently, $\phi M^{1/2} = 0$ at the theta temperature. To be strictly correct two diagrams should be presented: one operative for low-molecular-weight polymers in which entanglement is impossible $M < M_c$ and another which allows for the effects of entanglement. The usual representation (Figure 2) is applicable for $M > M_c$.

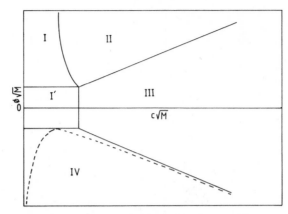

FIGURE 2. Phase diagram of polymer solution behavior obtained by the application of renormalization group theory. $\phi = (T - \theta)/\theta$. Dashed line represents the precipitation curve for the polymer.

Region I′: The polymer is sufficiently isolated for the condition $N\langle r^2\rangle^{3/2} < \beta$ to hold, where N is the number of chains per unit volume, $\langle r^2\rangle$ is the mean square end-to-end distance, and β is the excluded volume per statistical chain element.

Region I: The polymer molecule has a slightly expanded configuration, excluded-volume effects dominating in this region. The above region comprises the dilute solution region of Frisch and Simha.[3]

Region II: The intermediate concentration (semidilute and semiconcentrated) region, where polymer chains interpenetrate and is defined by the condition $N\langle r^2\rangle^{3/2} > \beta$. As in the Frisch–Simha approach, the lower limit is defined by the concentration at which two polymers overlap; this leads to the limits being $3M/4\pi N_A\langle S^2\rangle^{3/2} \leq c \leq M/N_A\langle S^2\rangle^{3/2}$, where N_A is Avogadro's number and $\langle S^2\rangle$ is the mean square radius of gyration. The upper limit corresponds to the dense polymer region where each monomer is in contact with many others, for example, Region III. It should be pointed out that the tie lines in Figure 2 do not imply that the end-to-end distance undergoes a discontinuous change but rather that a different scaling law applies. Interpolation equations have been derived which allow prediction of the end-to-end distance variation as a function of temperature and concentration.

Mean field[4] (MF) and scaling[5] (SL) theories lead to the prediction of a two-phase region associated with the precipitation of polymer below the theta temperature. The results of these theories are summarized in Table 1.

TABLE 1. Formulas for the Prediction of Regions of Polymer Solution Behavior
Illustrated in Figure 2

Theory	$\langle S^2 \rangle$		ξ^2	
	SL	MF	SL	MF
Region				
I′	n	n	—	—
I	$n^{6/5}\beta^{2/5}$	$n^{6/5}\beta^{2/5}$	—	—
II	$n\varrho^{-1/4}\beta^{1/4}$	$n\varrho^{-1/5}\beta^{1/5}$	$\varrho^{-3/2}\beta^{-1/2}$	$\varrho^{-6/5}\beta^{-4/5}$
III	n	n	ϱ^{-2}	

Recent neutron scattering measurements,[9-11] discussed in detail in Chapter
4, on polystyrene and polydeutrostyrene in cyclohexane have shown the
general validity of the ideas summarized above; however, these workers
indicate the possible existance of a further region in the semidilute-semi-
concentrated region probably reflecting the various stages in the modification of the hydrodynamic interaction terms as a consequence of polymer–
polymer interaction.

In this chapter, we are concerned with the effects of concentration on
the dynamic properties of the polymer which may also be expected to scale
as cM; however, as we shall see the operative function depends on the
wavelength-time scale of the interactions being probed. Before discussing
the effects of concentration it is worth considering the factors which influence the dynamic behavior of a polymer chain in dilute solution—Regions
I and I′.

2. DYNAMIC PROPERTIES OF A POLYMER CHAIN
IN DILUTE SOLUTION

The low-frequency spectrum of a polymer molecule in dilute solution
is dominated by the relaxation of the whole molecule either by overall
rotation or as a consequence of cooperative distortions, e.g., normal mode
relaxation (Figure 3). Inelastic collisions between the solvent and the polymer induce configurational changes which lead to an extension of the
polymer in the direction of shear stress. Application of high-frequency

oscillatory shear leads to the generation of waves with wavelengths shorter than the end-to-end distance for the polymer and the accompanying distortion will lead to nodes (X) within the polymer. As the frequency is increased, so the contribution of these higher modes decreases and this is observed as a reduction in the effective viscosity of the solution (Figure 3). The interaction between the solvent and the polymer chain is introduced via the Oseen hydrodynamic interaction parameter into the theories of normal mode relaxation.[1,2] From a molecular point of view a distortion of the overall polymer chain involves changes in *gauche–trans* ratio and excitation of torsional–librational motions. As such the excitation might be expected to be dependent upon the chemical structure of the polymer backbone; however, it is well established that the normal mode relaxation of a semi-flexible polymer is purely a function of the end-to-end distance and molecular weight. The obvious explanation is that the rate of conformational change is very rapid and hence has no effect on the determination of normal mode relaxation. The question does however arise as to whether the motions in the high-frequency tail of the normal mode spectrum and the primary rotational isomeric transitions are identical.

Data from three techniques will feature strongly in this chapter: nmr, dielectric, and ultrasonic relaxation. The first is indicative of the randomization of the magnetic polarization as a consequence of change; because

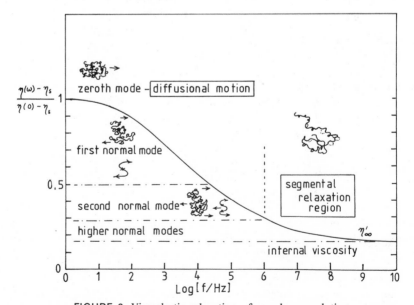

FIGURE 3. Viscoelastic relaxation of a polymer solution.

of the low natural abundance of ^{13}C nuclei it is possible, using this nucleus, to gain information on the motion of specific sites in the polymer. This technique has however the limitation of being a single-frequency observation and hence the interpretation of the observed relaxation times requires assumptions to be made with regard to the form of the distribution function applicable to the particular motion concerned. Dielectric relaxation monitors the rate of reorientation of specific molecular dipoles and as such can be used to provide information on specific types of molecular relaxation. Ultrasonic relaxation is the least specific of the three; however, it does provide an important link between these molecular probes and viscoelastic relaxation. The true behavior of a system can be only obtained by the rationalization of all experimental observations and it is therefore essential to understand the nature of the selection rules operative for a particular technique. Dielectric relaxation requires the rotation of a molecular dipole; NMR is sensitive to either the exchange of spin states or to randomization of the magnetic dipole vector. The first process dominates ^{1}H studies, whereas the latter operates almost exclusively for ^{13}C relaxation. Ultrasonic relaxation requires that energy originally in translational motion be converted and stored for a short time in some other form. For example, as a consequence of an inelastic collision a torsional oscillation in an energetically nondegenerate rotational isomer can be excited and lead to a transient increase in the population of the higher energy state. The selection rules for sound absorption are the occurrence of either an energy (ΔH^0) or volume (ΔV^0) difference between the states being perturbed. At low frequencies, all possible states can be excited and the energy loss will be large. Increasing the frequency will lead to a condition being reached whereby the rate of perturbation is faster than the excitation of the higher energy state and the sound absorption will be decreased. A typical plot for a solution of polystyrene in toluene is shown in Figure 4. The dispersion can be described in terms of an equation of the form

$$\frac{\alpha}{f^2} = A/[1 + (f/f_c)^2] + B \qquad (1)$$

where B is the high-frequency limiting value of the acoustic absorption, A is the amplitude of the relaxation process, and f_c is the characteristic frequency for the exchange process ($k_i = 1/2\pi f_c$). The magnitude of the sound absorption and the perturbation of the equilibrium can be derived using nonequilibrium thermodynamics.[12] The theory predicts that the quantity A is related to the incremental change in the compressibility $\mathscr{K}_s$, specific heat $-C_p$, volume ΔV^0, and enthalpy difference $-\Delta H^0$ be-

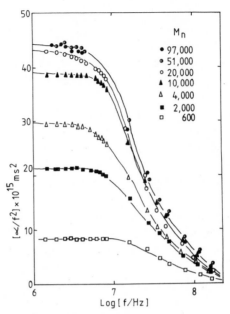

FIGURE 4. Ultrasonic relaxation of polystyrene in toluene.

tween states by

$$A = \frac{2\pi^2}{\varrho v C_p^0} \left[\left(\frac{\delta \mathscr{K}_s}{\mathscr{K}_s^0} \right) - \left(\frac{(\gamma - 1)\delta C_p}{C_p^0} \right)^{1/2} \right]^2 \qquad (2)$$

$$= \frac{2\pi^2}{\varrho v} \frac{(\gamma - 1)\delta C_p}{C_p^\infty} \left[1 - \frac{C_p^0 \Delta V}{V - \Delta H^0} \right] \qquad (3)$$

where v is the velocity of sound, $\tau = 1/k_i$, the relaxation time, and C_p^0 and C_p^∞ are the specific heats at constant pressure in, respectively, the low- and high-frequency limits. The value of C_p therefore corresponds to the situation where dynamic exchange between isomeric states cannot contribute due to the conformational exchange $\delta C_p = C_p^0 - C_p$. Similarly $\delta \mathscr{K}_s$ [$= 1/v(\partial V/\partial P)_s$] represents the difference in compressibility of the system allowing for and excluding the possibility of conformational change. The quantities ΔH^0 and ΔV^0 are, respectively, the enthalpy and volume change associated with the relaxation. If $\mathscr{K}_s$ or ΔV^0 are negligible then the relaxation amplitude is essentially proportional to the energy difference between states being perturbed. This is the condition which applies for most conformational changes. Alternatively, if δC_p and or ΔH^0 are zero the process is termed a volume relaxation; examples of this type of process would be hydrogen bond exchange and structural relaxation. For a monoatomic liquid, the value of B can be described by the "classical" Navier–Stokes equation:

$$B = 2\pi^2/\varrho v^3[(4/3\eta_s + \eta_v) + (\gamma - 1)Q/C_p] \qquad (4)$$

where η_s and η_v are, respectively, the shear and volume viscosities, ϱ is the density, γ is the ratio of the specific heat at constant pressure to that at constant volume, and Q is the thermal conductivity. For a simple liquid B is independent of frequency; however, for polymer solutions η_s should be replaced by $\eta_s(\omega)$; hence allow for the effects of a frequency dependence of the shear and normal mode relaxation. The theory of ultrasonic relaxation has been discussed elsewhere[12]; however, the above outline will serve to form the basis of a discussion for the relaxation behavior of polymer solutions.

3. MOLECULAR WEIGHT EFFECTS IN DILUTE SOLUTION

The nature of the molecular weight effect depends on the probe used in the investigations. A marked molecular weight dependence has been observed from dielectric measurements[13-22] (Figure 5). Increasing the molecular weight leads to a decrease in the relaxation frequency consistent with the thesis that the process controlling the reorientation of the dipole is rotational diffusion–zeroth normal mode motion (Figure 3). Above some

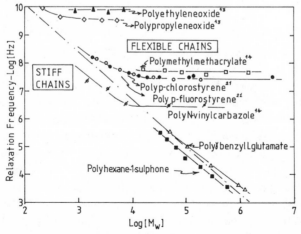

FIGURE 5. Effect of molecular weight on the dipole relaxation in various polymers.

critical molecular weight, the relaxation becomes independent of molecular weight. The point at which this change occurs depends on the chemical structure of the polymer. In the high-molecular-weight polymers, segmental motion occurs faster than whole molecule rotation and hence dominates the observed dielectric relaxation spectrum. Detailed investigations of halogenated polystyrene[21] have indicated that this thesis is correct. In the case of polypropyleneoxide, the stereochemical constraints on the possible relaxational motions lead to the observation of two relaxations: a high-frequency process associated with segmental motion and a lower-frequency process attributed to whole molecule, normal mode relaxation of the polymer. The dielectric data would imply that the relaxation spectrum is composed of these two regions; however, it could also be argued that the large dielectric loss observed at high frequency is purely the highest normal mode relaxations of the polymer.

Investigation of the ultrasonic relaxation of dilute solutions of narrow molecular weight distribution samples of polystyrene (Figure 4)[23,24] have indicated that the relaxation frequency exhibits a similar behavior to that observed in the dielectric studies mentioned above (Figure 3). However, in the case of ultrasonic relaxation the most marked effect is the change in amplitude of the relaxation with molecular weight. By analogy with the dielectric case[21] one would anticipate that the observed behavior was a consequence of differences in the effective contributions from whole molecule and segmental relaxation. Consideration of the processes involved indicates that "normal mode" relaxation can be estimated from a knowledge of the viscoelastic relaxation using Eq. (4). The difference between the experimental observed attenuation of the sound wave and the theoretical shear contribution due to segmental motion. Using the Wang and Zimm[6] theory to interpolate the viscoelastic data, the contribution in the megahertz frequency range can be calculated (Figure 6). A plot of the amplitude factor A, the relaxation frequency (f_c), and the contribution due to shear relaxation (shown dashed) as a function of molecular weight indicates that an alternative hypothesis is required to explain the acoustic data.

Studies have been performed for the separate stereoisomeric forms of the trimer[25] and tetramer[25] of polystyrene and these observations confirm that the process being observed is in fact a conformational change. The variation of f_c and A as a function of increasing molecular weight indicates that the major process being observed is one of conformational change; normal mode motions making only a minor contribution to the observed attenuation. It should be noted that in this polymer, the tail of the normal mode relaxation coincides with the "segmental" relaxation; however, its

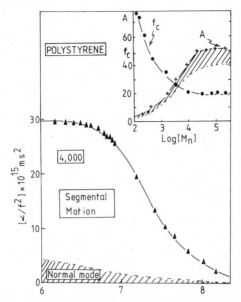

FIGURE 6. Ultrasonic relaxation in dilute polymer solutions. Insert is the variation of A, the acoustic amplitude, and f_c, the characteristic relaxation frequency with molecular weight.

amplitude, estimated from viscoelastic measurements, is too small to describe the observed ultrasonic attenuation.

The molecular weight dependence of the high-frequency relaxation in dilute solution is attributed to changes in the shape of the rotational potential controlling rotational isomerism with increase in chain length. In the oligomers, the number of nonbonding interactions acting on a particular subgroup of the molecule will depend very strongly on its position in the chain and as such the activation energy and energy difference between states (Figure 7). As the molecular weight is increased so the proportion of the polymer which experiences a potential profile which is "averaged" by being surrounded by other polymer elements is increased. At a molecular weight of approximately 10,000 the end groups have an insignificant effect on the relaxation and the observed acoustic dispersion is characteristic of local reorientational motions of the polymer backbone. The molecular weight dependence of the amplitude would also imply that short chains will have a different distribution of *gauche* and *trans* structures than those found in the higher-molecular-weight materials. Recent neutron scattering measurements[27,28] have indicated that the statistical distribution reflected in the end-to-end distance is different for short chains from long ones.

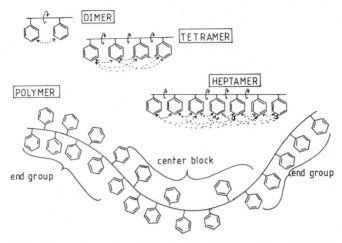

FIGURE 7. Change in nonbonded interactions with variation of chain length.

4. SIZE OF THE RELAXING SUBUNIT

The ultrasonic data would indicate that approximately six to eight monomer units are involved in the segmental relaxation process.[23,29,30] A variety of models have been proposed to analyze the ^{1}H and ^{13}C NMR relaxation for polystyrene solutions.[31-33] Extensive studies by Heatley and his co-workers[34-36] have indicated that the model proposed by Valeur,[33] which in the limit approximates to that which has been proposed by Jones and Stockmayer[37] is applicable to a large number of polymer systems. The basic model considers the polymer to approximate a diamond lattice and segmental rotation to involve three bond motions. The modified theory introduces two modifications to achieve a faster decay of the orientational autocorrelation function at long times. In the first modification, the correlation between conformational jumps in successive segments is cut off sharply rather than gradually along the chain and the second involves a relaxation of the strict lattice conditions allowing motions to occur even though neighboring bonds are not in precisely the correct conformations. Both modifications with suitable choice of parameters lead to autocorrelation functions $G(\tau)$ essentially identical to that suggested by Valuer *et al.*[33]:

$$G(\tau) = \exp(-\tau/\theta) \exp(\tau/\varrho) \operatorname{erfc}(\tau/\varrho)^{1/2} \qquad (5)$$

The correlation time ϱ characterizes three bond jumps, and the other θ

characterizes other processes. Analysis of data for polystyrene[31] using the above model has indicated that the data are best described in terms of coupled restricted rotations of segments involving between 9 and 13 bonds. The NMR data are influenced by the overall rotation of the polymer for molecular weights below 10,000. The autocorrelation function from this model is closely similar to Eq. (5) with $\tau/\theta = 1/12$ consistent with the results of Heatley and Wood[36] for polystyrene. NMR data therefore appear to imply that in many vinyl polymers the size of the relaxing element is of the order of ten monomer units.

An alternative method of estimating the range of nonbonded interactions with in a polymer chain is to systematically change the separation between interacting units (Figure 8). Ultrasonic[38] and ^{13}C NMR studies[39] of poly(styrene-alkane) and poly(α-methylstyrene-alkane) copolymers indicate that the activation energy for internal rotation of the styrene and alkane moieties are strongly coupled for short alkane blocks (Figure 8). Increasing the size of the alkane block leads to a situation in which the relaxation of the styrene and alkane moieties are resolved and occur essentially independently. Comparison of the activation energies of the styrene and α-methylstyrene copolymers with alkane blocks containing more than six bonds indicates a difference in activation energy of approximately 6 kJ/mol between the two series; the observed increase in the methyl-substituted polymers is consistent with the effects of nonbonded methyl interactions raising the barrier to internal rotation of the styrene moiety.[38] The obser-

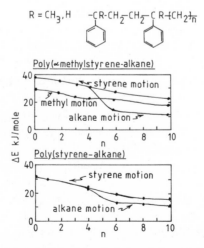

FIGURE 8. Activation energy versus alkane chain length for poly(α-methylstyrene-alkane) and poly(styrene-alkane) copolymers.

vation of an independence of the activation energy when the alkane block contains more than six carbon bonds indicates that the interaction between phenyl groups on neighboring styrene blocks are negligible. This observation would imply that in vinyl polymers coupling and interaction does in fact occur over as many as six bonds. This number is however rather smaller than implied from the analysis of the nmr data discussed above.

A recent investigation of the ultrasonic relaxation of polystyrene[40] and polyisobutylene[41] has indicated that at high temperatures the amplitude of the relaxation undergoes a rapid increase before once more undergoing the expected decrease. An analysis of the data is possible in terms of a rotational isomeric model in which at approximately 333 K certain of the elements of the polymer suddenly are able to contribute to the relaxation. Below 333 K only isotactic and atactic units appear to contribute to the observed relaxation; above 333 K the syndiotactic units contribute to the relaxation spectrum. A possible explanation of the observed behavior can be obtained by considering that the energy introduced by inelastic collisions at low temperature are only able to activate the most labile of torsional motions and these are associated with the isotactic and atactic units. At high temperature, energy fed into the isotactic and atactic units can also induce torsional oscillations in the neighboring syndiotactic elements and lead to facile internal rotation. Because the energy profile associated with internal rotation is the average of interactions extending over at least six units, introduction of activity in the syndiotactic dyads has little effect on the overall relaxation frequency; however, it does increase the number of monomer units contributing to the observed relaxation behavior and hence the amplitude of the acoustic attenuation. Evidence for this type of behavior has also been obtained from ^{13}C and ^{1}H NMR studies[42,44] and also light scattering experiments[45]; however, the effects observed are only just larger than the experimental error involved in the measurements. Recent neutron scattering measurements on partially deuterated polystyrene have indicated that as the size of the labeled block is shortened so the deviation from ideal Gaussian statistics increases.[27] This observation is consistent with the occurrence of short chain elements, typically of four to six monomer units in a pseudocoil form. It has also been observed that crystallization of polystyrene from solution leads to different structures dependent upon whether the experiment is performed above or below 340 K. There appears to be a significant body of evidence for the concept that local order may be induced as a consequence of intramolecular interactions.

5. SOLVENT EFFECTS

Investigations of the effect of change in the nature of the solvent have indicated that the activation energy for internal rotation is a function of solvent (Table 2). Using the simple rotational isomeric model it would be anticipated that the energy difference between states would be controlled by the local electric field.[46] For small molecules it has been found that the energy difference between states obeys an equation of the form[47]:

$$H_v^0 - H_l^0 = xh \qquad (6)$$

where $x = (\varepsilon - 1)/(2\varepsilon + 1)$, where ε is the dielectric constant of the liquid and $h = (\mu_A^2/a_A^3) - (\mu_B^2/a_B^3)$, μ_A and μ_B being the dipole moments of the states individved and a_A and a_B their molecular radii, and H_v^0 and H_l^0 are, respectively, the enthalpy differences in the vapor and liquid phase. This equation may be applied to the prediction of solvent effects by using either the theta condition or the vapor phase as reference.[41] Either approach does not allow prediction of the observed energy differences or the activation energy changes for internal rotation. The lack of agreement between theory and experiment can be ascribed to the effects of specific solvent interactions in the case of certain polymer–solvent combinations.[41]

It has however been pointed out that if it is assumed that the segment has not only to overcome the potential energy interaction resulting from nonbonded interactions between groups but also has to overcome the local friction coefficient[48], then subtraction of the viscosity should lead to a

TABLE 2. Effect of Solvent on the Activation Parameters for Segmental Motion

Polymer	Solvent	Activation energy (kJ mol⁻¹)		
		Effective	Viscous	Internal
Poly(methylmethacrylate)	Toluene	28.0	8.7	19.4
	Dioxane	30.1	13.0	20.0
	Tetraline	47.6	15.9	21.7
Poly(butylmethacrylate)	Toluene	22.6	8.7	14.4
	Dioxane	20.9	13.8	7.1
	Chloroform	15.0	6.7	8.3
	1,2-Dichloroethane	17.6	9.2	8.4

constant intramolecular contribution to the activation energy[49,50] (Table 2). From the evidence presented there would appear to be some basis for this proposition.

6. SEMICONCENTRATED SOLUTIONS

This region may be defined as lying between the limits of $[\eta]c \approx 1-4$. The principle effect of changes in the concentration of the polymer is to modify the viscoelastic–normal mode relaxation spectrum. These changes have been discussed elsewhere. From the above discussion it may be expected that the increase in concentration will lead to an increase in the magnitude of the relaxation associated with segmental relaxation and also lead to an increase in the normal mode relaxation contribution. Ultrasonic investigations[51] of the effects of changes in concentration on the relaxation spectrum for polystyrene in toluene solutions (Figure 9) indicates that up to approximately 10 wt % the observed relaxation is the sum of two contributions: normal mode and segmental processes. Estimation of the normal mode relaxation contribution using the Wang and Zimm theory[6] and adding a segmental contribution based on the dilute solution amplitude allows prediction of the observed dispersion characteristics within experimental error. This indicates that as far as the high-frequency behavior is concerned the introduction of hydrodynamic interactions and even increasing the polymer concentration to close to that of the volume filling limit has little effect on the short range dynamics. Clearly there will be a minor perturbation in the intramolecular energy profile consistent with the observed change in the coil dimensions; however, these effects appear to have little effect on the rate of segmental motion. Similar conclusions have drawn from dielectric[52] and NMR investigations[53] of solutions in this concentration range. These observations are consistent with the idea that the motion is influenced by the magnitude of the local friction coefficient. Although the zero frequency viscosity is markedly increased by the change in concentration, the effect of normal mode relaxation reduces the value of the friction coefficient to a value which is close to that of the solvent.

Investigations of the viscoelastic behavior of solutions in this region[54,55] have indicated that the ideal behavior has to be modified to allow for so-called "internal viscosity" effects (Table 3). The relaxation of the viscosity appears to reach a minimum value which is slightly higher than that of the solvent. It is not clear whether this residual increment relaxes out in the megahertz region or not.[56] It does however indicate that the effective friction

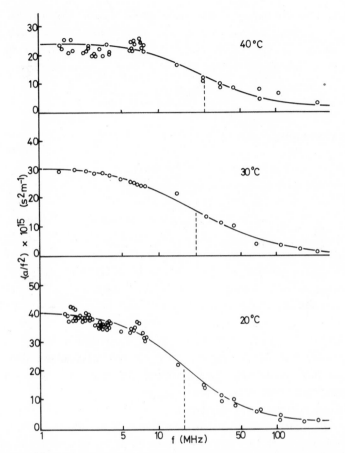

FIGURE 9. (a/f^2) vs. frequency for 2.6 wv % PS4000 in toluene: ultrasonic relaxation in semiconcentrated solutions of polystyrene.

coefficient controlling rotational isomerism is a function of the polymer solvent combination and implies that there may exist a case for invoking specific polymer–solvent interactions, albeit of a transient nature.

7. MODERATELY CONCENTRATED SOLUTIONS

A further increase in the concentration to a region where $[\eta]c$ lies between 4 and 10 leads to significant effects on the form of the high-frequency relaxation spectrum. The frequency of the relaxation is not very much affected by the increase in concentration. Dielectric studies[57] of polar

TABLE 3. Limiting Values of the Viscosity for Various Polymer Solutions[54]

Polymer–solvent	$M_w \times 10^3$	T (°K)	$10^3 [\eta'_\infty]$ (m³ kg⁻¹)^a
Polystyrene–DOP	98.2	283.2	15
	153	283.2	14
	411	283.2	15
Polystyrene–Arochlor	153	303.2	14
	411	303.2	14
	867	303.2	13
	2850	303.2	14
Poly(methylstyrene)–Arochlor	105	303.2	18
	386	303.2	18
	1400	303.2	19
Poly(1,4-butadiene)–Arochlor	92	303.2	8
	244	303.2	7
	910	303.2	8
Polyisobutylene–liquid paraffin	400	303.2	7

^{a} Error in η_∞ is approximately ± 2.

polymers in solutions, in which the viscosity has been enhanced by a non-polar polymer, indicate that little change in the relaxation behavior occurs on increasing the segment density.

Acoustic studies of polystyrene in toluene[51] indicate that while sub-traction of a contribution due to normal mode motion leads to a relaxation spectrum which is identical in shape to that observed in the semicon-centrated region, the amplitude of the relaxation is larger (Figure 10). This implies that either the number of relaxing segments has increased or that the energy parameters associated with the conformational change have been modified. Analysis of the temperature dependence of the acoustic relaxation amplitude indicates that the latter is the case. As indicated in the Intro-duction, the amplitude of the acoustic loss reflects the entropy, enthalpy, and free volume changes associated with the relaxation process. Investiga-tions of the concentration dependence of the adiabatic compressibility[51] indicates that the free volume changes only slightly as a consequence of the increase in concentration. The temperature variation of the acoustic at-tenuation indicates that the enthalpy difference is essentially unaltered from its dilute solution value. It is therefore concluded that the principal change which occurs is entropic in nature. By analogy with rubber elasticity theory, it can be proposed that this additional contribution is associated

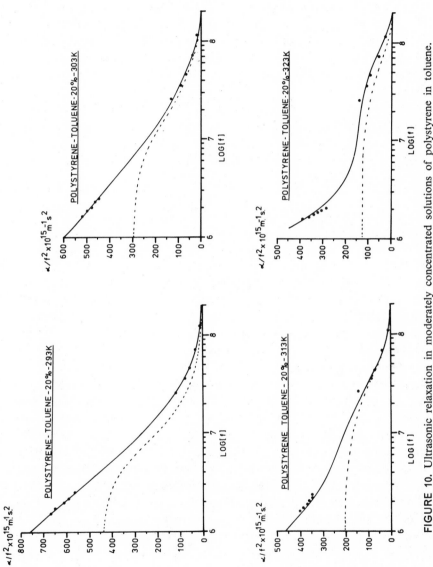

FIGURE 10. Ultrasonic relaxation in moderately concentrated solutions of polystyrene in toluene.

with the fluctuation induced in the "local" structure as a consequence of the stresses generated by the propagation of the sound wave. Although the solutions being investigated are well below the gel concentration, there is the possibility of forming for short times transient network structures; these at higher concentrations leading to stable gels.

The effects of polymer–polymer coil overlap have been reported from investigations of polyisobutylene in concentrated solutions in CS_2.[53] Increasing the concentration leads to the generation of a concentration region where the relaxation time is independent of further changes in segment density. The concentration independent region is also observed to be independent of molecular weight. An analysis of this data using the concept of a "blob" defining the range of interaction of a coil allows estimation of the number of such blobs per virtual chain. In the case of the polyisobutylene polymers the number is 15.

8. CONCENTRATED SOLUTIONS

At high concentrations polymer molecules undergo significant interpenetration and as a consequence will form a dynamic network structure. Because such systems lack chemical cross-links they are able to flow and hence behave as liquids. Viscoelastic investigations of the transition from concentrated solution to matrix-melt behavior are rare; however, the type of behavior observed parallels closely that found on changing the molecular weight of a polymer in the melt state. Low-molecular-weight short-chain materials behave very much like simple liquids. Increasing the molecular weight leads to typical Rouse-like behavior being detected and a dependence of the viscosity on the first power of the molecular weight. At a critical molecular weight M_c the chain is now sufficiently long to undergo interaction with its neighbors for a time which is comparable to that for normal mode relaxation. These polymer–polymer interactions impose on the normal mode relaxation spectrum a number of constraints and lead to the observation of a shifted contribution characteristic of the network formed. This so-called "terminal region" has recently been discussed theoretically in terms of reptation motion.[58-60] The theory predicts that in the region above M_c the viscosity should obey an M^3 dependence; however, experimentally an $M^{3.4-3.5}$ law is observed. The resulting relaxation spectrum is composed of two regions: the terminal region characteristic of the entangled network and a high-frequency spectrum defining the relaxation of chains between entanglements and similar to that predicted by Rouse

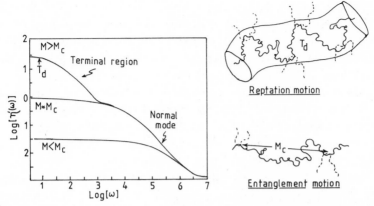

FIGURE 11. Dynamic viscosity as a function of molecular weight reptation motion.

theory for short chains, with odd modes suppressed (Figure 11). The visco-elastic behavior will be discussed elsewhere; however, it is interesting to see how the concepts of polymer–polymer interaction and entanglement can influence the high-frequency relaxation spectrum. A study of the high-frequency ultrasonic relaxation of polydimethylsiloxane[61] (Figure 12) has indicated that there must exist at least two relaxations in the region above 1 MHz. Comparison of the predicted attenuation using experimental measurements of the viscoelastic relaxation[62] in this region indicates that the dominant process in the 1–100 MHz region is normal mode relaxation. Investigation of the attenuation characteristics above 100 MHz indicates

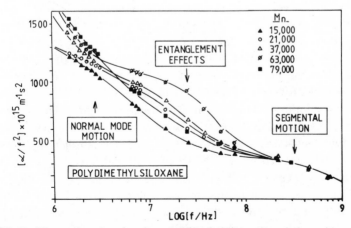

FIGURE 12. Ultrasonic relaxation in poly(dimethylsiloxane) variation with molecular weight.

that there is a further dispersion which is molecular weight independent and characteristic of segmental relaxation. Increasing the molecular weight leads to a deviation between the predictions from the viscoelastic experiments[62] and the observed acoustic attenuation.[61] The deviation reaches a maximum at approximately M_c and then decreases. This would imply that the difference between the observed ultrasonic attenuation and the shear contribution is associated with the effects of chain entanglement on the volume relaxation of the system. As in the discussion of polystyrene it is rather difficult to differentiate between volume and entropy contributions, the latter may be expected to be the more dominant.

The comparison of the ultrasonic and viscoelastic data, described above, indicates that significant polymer–polymer interactions occur and thus the approximation of an isolated chain moving in a rigid matrix may not be strictly correct. In the reptation model the motion is considered to be diffusional, in which case the investigation of the motion of a radio-actively tagged polymer is a very good analogy to the theoretical model.[63] Such studies indicate that the diffusion rate constant increases approximately as M for molecular weights below M_c. Attempts at diffusing high-molecular-weight materials have been rather unsuccessful; this is unfortunate since it is this region where the $M^{3-3.5}$ power should be observed. Below M_c the diffusion is observed to obey a relation of the type

$$\log(D) = Ad - (1/2.303f) \tag{7}$$

where f is the fractional free volume derived from the temperature dependence of the viscoelastic relaxation, which in turn was found to exhibit Rouse-like behavior. Proton NMR self-diffusion studies[64] of poly(ethylene-oxide) and polydimethylsiloxane performed over a concentration range from 0.5% to 100% polymer and a molecular weight range from 10^2 to 10^6 indicate that the self-diffusion depends both on concentration and molecular weight. In low-molecular-weight samples $\ln D$ is roughly proportional to concentration. In solutions containing high-molecular-weight polymer, plots of $\ln D$ versus concentration show an upward curvature. In the low-molecular-weight bulk polymer D_s varies as $M^{-1.7}$ and this parallels the observed variation of the viscosity. The NMR studies[64] indicate that D_s does not exhibit a marked change with increasing the molecular weight although the viscosity does. These data indicate that polymer entanglements influence self-diffusion and viscosity in different ways. Supporting this contention is the observation that the variation of the polymer self-diffusion in changing from a good to a bad solvent varies in a less pronounced manner than the change in the viscosity. These observations question the basis of the reptation concept; however, extension of the theory and analysis of the consequent data indicate that for certain NMR data[65] and viscoelastic data[66] the theory appears to describe the behavior adequately. Clearly the validification of the reptation concept will be one of the important objectives of polymer physics in the next few years.

In the context of the high-frequency relaxations, it is important to note that although the viscosity of the polydimethylsiloxane melts changed by several orders of magnitude the relaxation stayed essentially in the same position.[61] This observation indicates that despite that large changes in the low-frequency viscosity the local frictional interactions are essentially the same in the low- and high-molecular-weight materials. Comparison of the relaxation frequency obtained from different techniques indicates that the plot exhibits a marked elbow (Figure 13). This apparent deviation from simple activation theory can be explained by the use of "starvation kinetics." Eyring[67] has recently modified his original concepts to allow for the fact that the preexponential factor in the rate equation ($k = A \exp(-\Delta H^{\ddagger}/RT)$) is also temperature dependent. In the simple transition theory, it is assumed that there exists an equilibrium between molecules at the top of the barrier and the ground states. The exponential function defines the proportion of the total number of molecules which have sufficient energy to gain access to the transition state. In the case of polydimethylsiloxane,[61] the low value of $\Delta H^{\ddagger}$ leads to the prediction that all molecules are capable of undergoing a transition at room temperature. The rate of isomerization will then be controlled by the numbers that are activated by bimolecular collision—the temperature dependence of the preexponential factor.

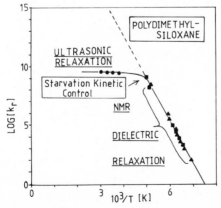

FIGURE 13. Activation energy plot for polydimethylsiloxane.

The effect of chain density on the behavior of the polymers in the concentrated region can be obtained from the effects of dilution on the ultrasonic relaxation spectrum. It is observed for low-molecular-weight polymers[68] that the addition of solvent leads to an immediate decrease in the observed attenuation. However, in the case of the polymers above M_c,

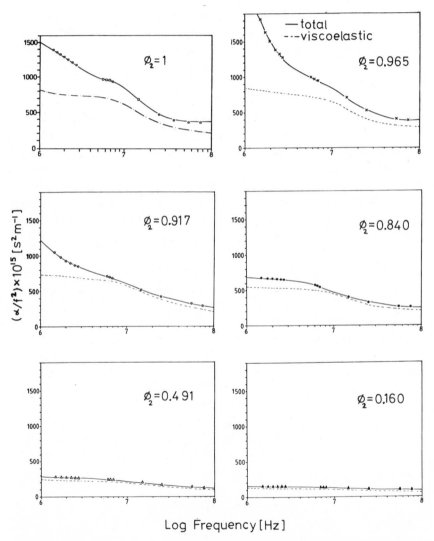

FIGURE 14. Absorption versus frequency for PDMS ($\bar{M}_n = 7.94 \times 10^4$) in toluene at 303 K: effect of concentration on the ultrasonic relaxation spectrum for poly(dimethylsiloxane).

addition of solvent leads to an increase in the acoustic attenuation despite the observation that the viscosity changes very little (Figure 14). The introduction of solvent into a physically entangled matrix will lead to a swelling of the polymer chains and hence an increase in the number of chain–chain interactions.[69,68] The natural effect of the addition of solvent would be to decrease the number of coil–coil contacts and hence lower the viscosity, this is contrary to observation.[69] The increase in the acoustic attenuation is also consistent with there being a greater probability of polymer–polymer contacts and reflects the growth rather than depletion of the interactions. A further increase in the amount of solvent leads to a decrease in the observed viscosity and acoustic attenuation consistent with there being a reduction in the number of chain–chain contacts, e.g., entanglements. Further dilution of the solution leads to the observation that the ultrasonic relaxation conforms to a simple viscoelastic model at a point corresponding to $c[\eta] = 10$ consistent with the concepts outlined above. It is not at present possible to analyze this data further because of a lack of precise viscoelastic data for this type of system. Corresponding measurements on the ultrasonic behavior of the slightly more hindered polymethylphenyl-siloxane[70] confirm the interpretation of the behavior observed in poly-dimethylsiloxane.

9. CONCLUSION

The above review has attempted to illustrate the factors which influence the high-frequency dynamic behavior in polymers both in solution and in their melt phase. It is by no means comprehensive and the examples chosen have been selected to illustrate a point and may not be the most definitive examples of a particular type of behavior in the literature. Discussions of the way in which high-frequency melt behavior relates to relaxation in the solids and also high-frequency relaxation in hydrogen-bonded and interacting polymer systems will be discussed elsewhere.[71] Likewise the reader interested in the discussion of dilute solution relaxation behavior of polymers or of the detailed discussion of acoustic relaxation is referred to other review articles on these topics.[12,72]

In summary, it is becoming clear that the behavior of polymer molecules in solution depends not only on the end-to-end distance but also on the detailed chemical structure of the polymer. Which of these factors is more important will depend upon the time scale of the observation and also on the type of probe used.[73,74] It also appears relatively well established that normal-mode–coherent distortions of long sections of polymer chain can be differentiated from local-segmental, incoherent distortions of local segments of the polymer. In the high-frequency region, both normal mode and segmental relaxations are influenced by the chemical structure of the backbone, however, in slightly different ways. Increases in the polymer chain density do not have the marked effects they have at low frequency; however, in the correct molecular weight time domain the processes of chain entanglement can have a very significant effect on the observed relaxation behavior.

REFERENCES

1. P. A. Rouse, *J. Chem. Phys.* **21**, 1272 (1953).
2. B. H. Zimm, *J. Chem. Phys.* **24**, 269 (1956).
3. R. Simha and L. Ultracki, *J. Polym. Sci.* **A2** (5), 853 (1967).
4. S. F. Edwards, *Proc. Phys. Soc. (London)* **88**, 265 (1966).
5. P. G. de Gennes, *Macromolecules* **9**, 587 (1976).
6. F. G. Wang and B. H. Zimm, *J. Polym. Sci. Polym. Phys.* **12**, 1619 (1974).
7. M. Daoud, J. P. Cotton, Farnoux B., G. Jannink, G. Sarma, H. Benoit, R. Duplessix, C. Picot, and P. G. de Gennes, *Macromolecules* **8**, 804 (1975).
8. S. F. Edwards and E. F. Jeffers, *J. Chem. Soc. Faraday Trans 2* **75**, 1020 (1979).
9. R. W. Richards, A. Maconnachie, and G. Allen, *Polymer* **19**, 266 (1978).
10. J. P. Cotton, M. Nierlich, F. Boue, M. Daoud, B. Farnoux, G. Jannink, R. Duplessix, and C. Picot, *J. Chem. Phys.* **65**, 1101 (1976).
11. R. W. Richards, A. Maconnachie, and G. Allen, *Polymer* **22**, 147 (1981).
12. E. Wyn-Jones, and R. A. Pethrick, *Topics in Stereochemistry* (E. L. Eliel and N. Allinger, eds.), Vol 5, p. 205, Wiley-Interscience, New York (1970).
13. W. H. Stockmayer (1967), *Pure Appl. Chem.* **15**, 539 (1967).
14. A. M. North and P. J. Phillips, *Chem. Commun.* **1340** (1968).
15. K. Matsuo, Ph. D. thesis, MIT (1960).
16. H. Block, E. F. Hayes, and A. M. North, *Trans. Faraday Soc.* **66**, 1095 (1970).
17. T. W. Bates, K. J. Ivin, and G. Williams, *Trans. Faraday Soc.* **63**, 1976 (1967).
18. S. Mashimo, *Macromolecules* **9**, 91 (1976).
19. C. H. Porter, J. H. Lawler, and R. H. Boyd, *Macromolecules* **3**, 308 (1970).
20. C. H. Porter and R. H. Boyd, *Macromolecules* **4**, 589 (1971).
21. W. H. Stockmayer and K. Matsuo, *Macromolecules* **5**, 766 (1972).
22. W. H. Stockmayer, A. A. Jones, and T. L. Treadwell, *Macromolecules* **10**, 762 (1977).
23. M. A. Cochran, A. M. North, and R. A. Pethrick, *J. Chem. Soc. Faraday Trans. 2* **70**, 215 (1974).
24. H. Nomura, S. Kato, and Y. Miyahara, *Nippon Kagahu Zasshi* **88**, 502 (1967).
25. B. Froelich, B. Jasse, and L. Monnerie, *Chem. Phys. Lett.* **44**, 159 (1976).
26. B. Froelich, B. Jasse, C. Noel, and L. Monnerie, *J. Chem. Soc. Faraday Trans. 2* **74**, 445 (1978).
27. R. Duplessix, J. P. Cotton, H. Benoit, and C. Picot, *Polymer* **20**, 1181 (1979).

28. F. Boue, Diplome de Docteur These Université Louis Pasteur, Strasbourg (1977).
29. J. H. Dunbar, A. M. North, R. A. Pethrick, and D. B. Steinhauer, *J. Chem. Soc. Faraday 2* **70**, 1478 (1974).
30. M.A. Cochran, P. B. Jones, A. M. North, and R. A. Pethrick, *Trans. Faraday Soc.* **68**, 1719 (1972).
31. F. Laupretre, C. Noel, and L. Monnerie, *J. Polym. Sci. Polym. Phys.* **15**, 2127 (1977).
32. B. Valeur, L. Monnerie, and J. P. Jarry, *J. Polym. Sci. Polym. Phys.* **13**, 675 (1975).
33. B. Valuer, J. P. Jarry, F. Geny, and L. Monnerie, *J. Polym. Sci. Polym. Phys.* **13**, 2251 (1975).
34. F. Heatley and M. K. Cox, *Polymer* **18**, 225 (1977).
35. F. Heatley, A. Begum, and M. K. Cox, *Polymer* **18**, 637 (1977).
36. F. Heatley and B. Wood, *Polymer* **19**, 1405 (1978).
37. A. A. Jones, and W. H. Stockmayer, *J. Polym. Sci. Polym. Phys.* **15**, 847 (1977).
38. A. M. North, R. A. Pethrick, and I. Rhoney, *J. Chem. Soc. Faraday Trans. 2* **70**, 223 (1974).
39. A. V. Cunliffe and R. A. Pethrick, *Polymer* **21**, 1025 (1980).
40. J. H. Dunbar, R. A. Pethrick, A. M. North, and B. T. Poh, *Polymer* **21**, 764 (1980).
41. A. M. North, R. A. Pethrick, and B. T. Poh, *Polymer* **21**, 772 (1980).
42. K. J. Liu and R. Ullman, *Polymer* **6**, 100 (1965).
43. M. Kobayaski, K. Tsumura, and H. Tadokora, *J. Polym. Sci.* **A2** (6), 1492 (1968).
44. W. R. Krigbaum, F. Mark, J. G. Pritchard, W. L. Hunter, and A. Ciferri, *Makromol. Chem.* **65**, 101 (1963).
45. C. Reiss and H. Benoit (1961). *C. R. Acad. Sci.* **253**, 268 (1961).
46. R. J. Abraham and E. Bretschneider, *Internal Rotation in Molecules* (W. J. Orville Thomas, ed.), p. 481, Wiley, London (1974).
47. R. J. Abrahams, K. J. Pacjler, and P. L. Z. Weiss, *Phys. Chem.* **58**, 257 (1968).
48. A. T. Bullock and G. G. Cameron, *Structural Studies of Macromolecules by Spectroscopic Methods* (K. J. Ivin, ed.), p. 273, Wiley, New York (1976).
49. E. Helfand, *J. Chem. Phys.* **54**, 4651 (1971).
50. A. T. Bullock, G. G. Cameron, and P. M. Smith (1974). *J. Chem. Soc. Faraday Trans. 2* **70**, 1202 (1974).
51. J. I. Dunbar, A. M. North, R. A. Pethrick, and D. B. Steinhauer, *J. Polym. Sci. Polym. Phys.* **15**, 263 (1977).
52. A. M. North, *Chem. Soc. Rev.* **1**, 49 (1972).
53. J. P. Cohen-Added, A. Guillermo, and J. P. Messa, *Polymer* **20**, 536 (1979).
54. A. J. Matheson and B. J. Cooke, *J. Chem. Soc. Faraday Trans. 2* **72**, 679 (1976).
55. A. J. Matheson and B. J. Cooke, *Faraday Symp. Chem. Soc.* **6**, 194 (1972).
56. K. Osaki, *Adv. Polym. Sci.* **12**, 1 (1978).
57. W. N. Hunter and A. M. North, *Polymer* **20**, 1179 (1979).
58. P. G. de Gennes and L. A. Lazar, *Rev. Phys. Chem.* **33**, 49 (1982).
59. M. Doi and S. F. Edwards, *J. Chem. Soc. Faraday Trans. 2* **74**, 1818 (1978).
60. M. Doi (1980). *J. Polym. Sci. Polym. Phys.* **18**, 1005 (1980).
61. W. Bell, A. M. North, R. A. Pethrick, and B. T. Poh, *J. Chem. Soc. Faraday Trans. 2* **75**, 1115 (1979).
62. A. J. Barlow, G. Harrison, and J. Lamb, *Proc. R. Soc. (London) Ser. A* **282**, 228 (1964).
63. S. P. Chen and J. D. Ferry, *Macromolecules* **1**, 270 (1968).

266 CHAPTER 5

64. J. E. Tanner, K. J. Liu, and J. E. Anderson, *Macromolecules* **4**, 586 (1971).
65. R. Kimmich, *Polymer* **18**, 233 (1977).
66. W. Grassley, *J. Polym. Sci. Polym. Phys.* **18**, 27 (1980).
67. H. Eyring, *Science* **199**, 740 (1978).
68. W. Bell and R. A. Pethrick, *Acoust. Lett.* **3**, 156 (1980).
69. W. Bell, J. Daly, A. M. North, R. A. Pethrick, and B. T. Poh, *J. Chem. Soc. Faraday Trans. 2* **75**, 1452 (1979).
70. A. R. Eastwood, A. M. North, and R. A. Pethrick, unpublished data.
71. R. A. Pethrick, *Aggregation Processes in Solution* (E. Wyn Jones and J. Gormally, eds.), Elsevier, New York, Chapter 20 (1983).
72. R. A. Pethrick, *J. Macromol. Sci. Rev. Macromol. Chem.* **C9**, 91 (1973).
73. D. Pugh and R. A. Pethrick, *Chem. Britain* **70** (1980).
74. D. A. Jones, M. Lopez de Haro, and D. Pugh, *J. Polym. Sci. Polym. Phys.* **16**, 2215 (1975).

6

Pulse-Induced Critical Scattering (PICS), Its Methods and Its Role in Characterization

H. GALINA, M. GORDON, B. W. READY, and L. A. KLEINTJENS

1. INTRODUCTION

Pulse-induced critical scattering, or PICS,[1,2] is a technique for exploring the onset of phase separation in polymer systems. It has been developed at the Institute of Polymer Science of Essex University during the 1970s, with much cooperation from Koningsveld and his co-workers at the Central Laboratory of the DSM Company in Holland. The intensity of light scattered from a laser beam at selected angles is used to monitor the response of small samples to periodic fast thermal steps or "pulses." Much sensitive information is derived from the results, whose significance is explained below by reference to the typical and schematic *phase diagram* in

H. GALINA ● Instytut Technologii Organiccznej, Wroclaw, Poland.   M. GORDON ● The Statistical Laboratory, University of Cambridge, Cambridge CB2 1SB, United Kingdom.   B. W. READY ● Department of Chemistry, Wivenhoe Park, Colchester, United Kingdom.   L. A. KLEINTJENS ● Research and Patents, DSM, Geleen, Holland.

Figure 1. The purpose of PICS experiments includes the measurement of the lines and special points in such phase diagrams.

Despite the relative simplicity of operation (Section 4) of the PICS instrument described in Section 4.2, which has allowed it to be rather fully automated, its use extends beyond that of an exploratory tool for surveying miscibility or *compatibility* in polymer systems, though it does fulfil this role over wide temperature ranges (up to about 300 °C). The PICS method is fundamentally well placed for *characterization of polymers.* After suitable calibration, the sensitivity of phase changes to many structural variables can be exploited for characterizing, e.g., copolymer composition, tacticity, or cross-linking.

Characterization generally begins with molecular weight and its distribution, more specifically with polydispersity as expressed by functions of the moments of the molecular weight distribution, and this theme has

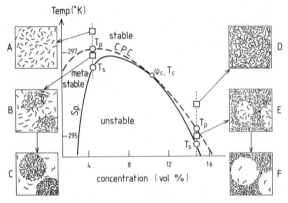

FIGURE 1. Typical phase diagram for polydisperse polymer solution, with schematic molecular interpretations. Plots of spinodal (solid curve Sp) and cloud-point curve (broken curve CPC) in the plane of temperature volume-per-cent concentration of polymer. A PICS experiment uses thermal pulses along the vertical broken lines shown for a dilute solution on the left and a concentrated one on the right. A pulse steps the temperature from the fixed upper square in the region of stable solutions to the lower square here shown in the metastable region, and back to the upper square. [The next pulse the sample undergoes will displace the lower square by $\sim 0.03°$ down towards the spinodal temperature T_s (cf. text of Section 4.1) leading to increased light scattering (Figure 6).] The critical point is denoted ϕ_c, T_c. The spinodal and cloud-point curves exemplify computer calculations for the free-energy of mixing functions of Eq. (4), with parameters shown in Table 1 for polystyrene/cyclohexane, viz., for a specific hypothetical mixture of three sharp fractions. The mixture comprises 6.5% of $M = 629$, 63.9% of $M = 229$, and 29.6% of $M = 42$ kg/mol. This mixture agrees in M_w, M_z, M_{z+1}, M_{z+2}, and M_{z+3} with the Flory distribution of $M_w = 200$ kg/mol, and thus simulates its phase diagram rather closely.

been amply demonstrated quite recently.[3,4] We update the characterization of linear polystyrene (PS) in Section 5 with some new data and a revision of parameters. Our main purpose is to present an account of the principles of operation of PICS and document their implementation (Section 4) with the Mark III instrument which is described in Section 4.2. First, however, the scientific theories, on which a proper understanding of the method rests, are presented with the minimum of mathematics. Incidentally, we illustrate the potentialities of the method with recent data obtained for a variety of polymer systems (see Figures 4, 5, and 12).

The use of PICS for basic characterization of polymers leans on theories whose chief ingredients, for our purpose, will be briefly reviewed under two headings. These are as follows:

(a) The classical thermodynamic theory (Section 2 and Appendix) of phase equilibria, with the theoretical *constraints* it imposes *on the form* of boundary lines or surfaces which separate the phases in the familiar phase diagrams. Thermodynamic theories are, of course, purely phenomenological in nature.

(b) Simple aspects of the thermodynamic theory of light-scattering (Section 3) developed by Debye[5] for systems approaching a critical point, or more generally a spinodal boundary (Scholte[6,7]).

The more intimate modern theories of statistical mechanics, starting in polymer science with the Flory[8]–Huggins[9] theory in 1941, which *quantitatively calculate* phase boundaries using a minimum of adjustable parameters have been expounded recently elsewhere.[3]

Thus in Section 2 we survey the shapes of relevant phase diagrams (Figure 1). The underlying thermodynamic theory of gross constraints on the shapes of such diagrams necessarily apply universally, to dilute solutions or just as well to bulk blends. We also mention the more intimate microscopic phenomena, namely, fluctuations[5] in density and especially those in concentration, which occur in neighborhoods of the boundary lines shown in Figure 1, and which, through the irregular inhomogeneities in refractive index which they induce, lead to dramatic changes in low-angle light scattering. Such *fluctuation-scattering* serves to locate the so-called *spinodal*[10] *curves* in the phase diagrams by means of PICS, using the Debye–Scholte extrapolation procedure (Section 4). Besides fluctuation phenomena, one may also observe the formation or existence of stable spherical particles in certain regions of the phase diagram, and light scattering from such particles or droplets serves to locate the *cloud-point curves* in the phase diagram.

2. DISCUSSION OF THE PHASE DIAGRAM

Figure 1 displays the usual curves observed in phase diagrams, in the composition–temperature plane, in the vicinity of phase separations occurring in mixtures. The so-called *cloud-point curve* CPC separates the stable from the metastable region in which, after some delay due to the need for a separate phase to be nucleated, two-phase emulsions are formed. At sufficiently low viscosities the final state shows gross separation of two phases when the droplets have coalesced and settled. The *spinodal curve* Sp separates the metastable from the unstable region, in which no energy barrier to be overcome in nucleation processes opposes the separation of the phases. As a result, in liquids of ordinary viscosity, spinodal decomposition occurs practically instantaneously in the unstable concentration–temperature domain. The molecular structure of the different states is schematized in three panels at each side of the figure. The onset of phase separation is foreshadowed by substantial concentration fluctuations as shown in the middle panel on each side. The underlying cause is the approach of the spinodal where the restoring force against a chance concentration fluctuation during the thermal translational (diffusion) motions vanishes. Fluctuations also occur above the cloud point, but these are generally strong only in samples at or close to the critical concentration ϕ_c, and very weak at concentrations widely different from ϕ_c. Fluctuations become strong at all concentrations inside the metastable region (Figure 1) as we approach the spinodal temperature, where the irregular fluctuation region would indeed grow to infinite size, were it not cut short when the critical size of a nucleus becomes exceeded. When this happens, it acquires a regular spherical shape (panels C and F, Figure 1) and proceeds rapidly to grow into a finite droplet or particle of the two-phase emulsion, which is stable except for a tendency of slow coalescence of droplets. Concentration fluctuations constitute inhomogeneities of the refractive index of a mixture, and these are detectable by low-angle light scattering.

As thermodynamics requires, the cloud-point and spinodal curves are seen to touch and share a common tangent at the so-called *critical point*. This point does *not* generally lie at the maximum of the curve for the upper critical solution point (UCST), or at the minimum for the lower critical solution point (LCST); only in cases of strictly binary mixtures, e.g., homodisperse polymer and solvent, does the critical point coincide with an extremum. For a mixture of more than three components, even for a solution of a very sharp polymer fraction with a narrow distribution of molecular weights, the critical point moves appreciably away from the

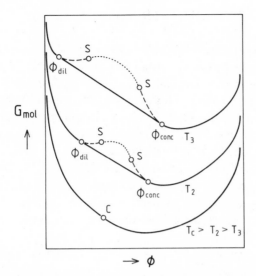

FIGURE 2. Schematic diagram for the molar free energy of a polymer solution in dependence on the volume fraction of polymer. The bottom curve shows the critical point C where the radius of curvature $= \infty$. At lower temperatures (higher G_{mol}) an indentation is seen to have grown out from that point, shown by the broken portions (metastable regions) and dotted portions (unstable regions) of the two sample curves presented. $S =$ spinodal points (inflection points), ϕ_{dil} and ϕ_{conc} denote the volume fractions of polymer in the dilute and concentrated solutions, respectively, which are in equilibrium at the temperature concerned.

extremum of the curve, and for polymer solutions in low-molecular-weight solvent it moves to some higher polymer concentration, as shown.

The critical scattering phenomenon is due to the metastable inhomogeneities in refractive index induced by the irregular concentration fluctuations just described, which foreshadow the approach to phase separation. The relevant theory, developed by Debye *et al.*,[11,12] is a thermodynamic one. As a first approximation, and for a binary mixture, the theory makes the scattered intensity I_0, in the limit of zero scattering angle θ, inversely proportional to the *curvature* of the plot of molar free energy against concentration. Three typical plots of this kind are shown in Figure 2. The two upper plots (corresponding to lower temperatures) each show two inflection points marked S, to signify spinodal points. Clearly the curvature of the plot vanishes at such a point, so that I_0 will, in the first approximation, become infinitely large there. The reason is easy to understand. The explanation, together with further thermodynamic implications of Figure 2, is given in the Appendix.

3. CRITICAL LIGHT SCATTERING

Scholte has used a standard light-scattering instrument, and Chu and co-workers[13] used specially constructed and highly sensitive equipment for locating spinodals (which they call pseudo-spinodals) by extrapolation of scattering intensities confined to stable samples (in the region S). The latter workers used a generalized nonlinear extrapolation method, but the classical linear formula of Eq. (3) below has been found applicable to PICS data. The PICS instrument (Section 4.2) is able to penetrate into the metastable region M, and thus to approach more closely to the spinodal; it makes measurements more quickly and conveniently, and generally with a substantial gain in accuracy as regards locating spinodal temperatures, except at concentrations remote from the critical.

3.1. Quantitative Theory

The intensity I of light scattered at angle θ to the incident beam is given, by the thermodynamic theory, in terms of the free energy ΔG of mixing and concentrations (conveniently taken as volume fractions) ϕ_i of the ith polymer species in the molecular distribution, by the proportionality relation:

$$I \propto TP(\theta)/|\,\delta^2\Delta G/\delta\phi_i\delta\phi_j\,| \qquad (i, j = 1, 2, \ldots) \tag{1}$$

The determinant in the denominator has been written out in Eq. (A1) of the Appendix, where we explain its role as a measure of the restoring force against the creation of a concentration–fluctuation. $P(\theta)$ is the properly averaged particle scattering factor. This factor, which is of course of primary importance in studies of the size and shape of particles (for an example see Burchard and Kajiwara[14]), is merely a constant in the intensities measured by PICS at constant θ, and can be absorbed into the unimportant proportionality constant implied by $\propto$. Near the spinodal point, where the spinodal determinant in the denominator of Eq. (1) vanishes, the Debye–Scholte theory expands the denominator in a Taylor series, truncated after the linear term, in powers of the distance $\delta T = T - T_s$ from the spinodal temperature. The constant term is then proportional to $\sin^2(\theta/2)$:

$$I \propto T/[c_0 \sin^2(\theta/2) + c_1\delta T] \tag{2}$$

In the range $0.03° < \Delta T < 5°$, and an angle of θ of about $30°$, the equation is approximated by

$$I \propto 1/\delta T \tag{3}$$

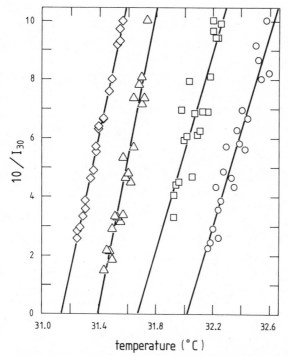

FIGURE 3. Typical Debye–Scholte plots (computer-drawn regression lines) for four solutions of polystyrene sample 39 (cf. Table 2 and Figures 10 and 11) in cyclohexane, obtained in one PICS run. Abscissa: temperature; ordinate: proportional to reciprocal intensity of scattering at 30°. This sample, a Flory (most probable) distribution of M_w $= 4.8 \times 10^6$: $\diamondsuit$, $\phi = 0.0573$, $T_s = 31.135 \pm 0.017\,°C$; $\triangle$, $\phi = 0.0372$, $T_s = 31.391$ $\pm\ 0.060\,°C$; $\square$, $\phi = 0.0245$, $T_s = 31.674 \pm 0.115\,°C$; $\bigcirc$, $\phi = 0.0212$, $T_s = 32.019\,°C$. Note the increase in experimental scatter and in the resulting inaccuracy of the extrapolation to T_s (intercept on abscissa) with decreasing concentration ϕ.

This theoretical prediction is verified below by the linearity of Debye–Scholte plots of $1/I_{30}$ against T (Figures 3–5).

4. PICS TECHNIQUE FOR LOCATING SPINODALS

4.1. Principle of Collecting Data

Pairs of data points (I_{30}, T) are measured and collected by the PICS instrument for the subsequent computerized production of Debye–Scholte plots (Figures 3–5). Their extrapolation leads to data pairs (ϕ_s, T_s), from which in turn the phase diagram is built up (see Figure 10, later). The *prin-*

ciple of obtaining the data points has not changed since the simple Mark I PICS instrument was described.[1,15] The more advanced design of Mark II was outlined in a latter publication,[16] and in Section 4.2 the fully computerized Mark III is illustrated and its operation is described.

The sample of total weight $\sim$12–15 mg is contained in a thin-walled glass capillary cell of 1 mm inner diameter. It is submitted to a series of thermal pulses; each pulse takes the sample from a very stable state (e.g., 20° above the cloud point curve) to a temperature in or near the metastable region in one fast step, and back to the original temperature in a second fast step. Between the two steps of each pulse, the sample is held at constant temperature T; successive pulses change T typically by $\sim$0.03° in the direction of the spinodal of the sample. The holding time at T is at least about 10 s for the case of solutions of polymers in solvents, but can be as high as 30 min for very viscous bulk mixtures. The holding time

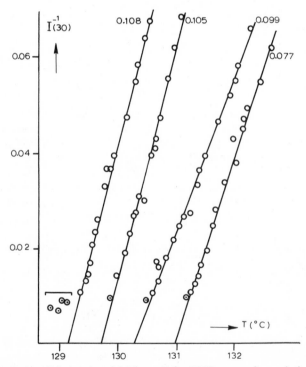

FIGURE 4. Debye–Scholte plots (cf. Figure 3) for PICS run on four solutions (at volume fractions indicated) of linear polyethylene ($M_w = 70$ kg/mol). The dotted circles at the end of the run (cf. text) were omitted from the regression line computation. Note the sensitivity to polymer volume concentrations and the precision of these measurements at elevated temperature. Solvent: diphenyl ether.

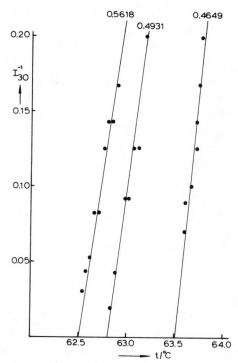

FIGURE 5. Debye–Scholte plots (cf. Figure 3) for three mixtures of low-molecular-weight polystyrene ($M_w = 2.10$ kg/mol) and polyisoprene ($M_w = 0.82$ kg/mol). The parameter values are weight fraction concentrations.

should be set by the operator so as to be long enough to allow the meta-stable fluctuation equilibrium to be established, but not so long as to allow emulsion particles to form and grow to a size too large for their rapid dissolution on heating. For solutions, the fluctuation equilibrium is set up faster than thermal equilibrium, which takes ~ 7 s to reach $\Delta T \sim 0.01°$. Scattering intensities at $30°$ and $90°$ are continuously recorded as shown in Figure 6 for four cells undergoing the pulses in rotation; I_{90} and I_{30} values are also sampled at the proper times electronically and processed by computer as described in Section 4.23.

The light intensities recorded in Figure 6 refer to four sample cells, and a total temperature range of $\sim 0.4°$ is covered (linearly decreasing temperature from right to left, total time ~ 6 min). The cells are labeled 1–4 in order of increasing spinodal temperature. We see that cell 1 shows practically no scattering ($I_{30} = I_{90} = 0$), while cell 2 is approaching sufficiently closely to its T_s, to give measurable scattering intensities, which are increasing slowly as T falls from right to left. The same behavior is more marked in cell 3, while cell 4 shows the rapid rise in successive pulses which is typical for a temperature range very close to T_s. The cells 3 and 4 show that about the first half of each pulse covers the rise in scattering intensity associated with the delay in thermal equilibration. The second (left-hand) half of each I_{30} pulse of cell 4 shows a slow rise, with a little noise superposed on the signal. The slow-rise portions of the I_{30} pulses of cells 3 and 4 can be joined up to exhibit their congruence with a single smooth intensity–temperature curve (shown as a broken line for cell 3 only) except for the significant deviation at the pulse marked X of cell 4 (see below). This broken curve represents the temperature dependence of scattering intensity (at 30°) under metastable fluctuation equilibrium. (The need to sample this curve *intermittently* may be understood by comparison with the intermittent use of metastable conditions elsewhere, e.g., in the nuclear scientist's bubble chamber.) The fact that after thermal equilibration, each pulse profile[1] conforms with the others to a smooth curve is proof that the observed scattering is indeed due to fluctuation equilibrium, and not significantly affected by phase separation. While the first five I_{30} pulses (counting from the right) of cell 4 approach closely to such a smooth curve, the left-most pulse shown for cell 4, taken at a temperature close to its T_s, shows a maximum at X in its profile, characteristic for onset of separation droplets, as schematized in panels C and F of Figure 1. The

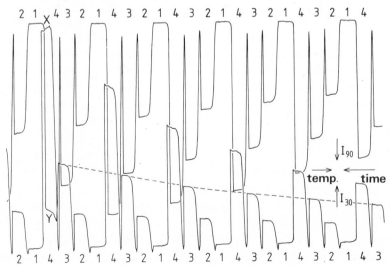

FIGURE 6. Recorder plot—to be read from right to left—of a PICS run with four sample tubes (numbered 1–4). For details see text.

maximum is generally much sharper in the profile of the I_{90} scattering pulse, as is also plain when we compare the left-most I_{30} and I_{90} pulse records in Figure 6 at X and Y, respectively. I_{30} intensity data must be excluded from the Debye–Scholte extrapolation plot if they are thus affected by phase separation. The I_{90} data are used in measuring spinodals only as a sensitive test for locating the temperature limit beyond which I_{30} pulse data must be rejected. I_{90} is usefully employed in addition for locating cloud points with the seeded-emulsion technique described earlier.[1] One advantage of this technique lies in the avoidance of the supercooling error[1] customarily made in the conventional approach to the cloud point by cooling, even when rather slow cooling rates are employed.[17] The operational procedure is outlined at the end of Section 4.23.

Apart from the phase-separation effects upon spinodal measurements (e.g., at X; Figure 6) there are other effects which will prevent observation of indefinitely large I_{30} values as T_s is approached. The Debye theory expressed in Eq. (2) contains a cutoff effect in the term $c_0 \sin^2(\theta/2)$ in the denominator, which takes effect at finite scattering angles θ. At spinodal points very close to the critical, and at temperatures $T - T_s < 0.03°$, we must expect this term to limit I_{30} and thus to raise $1/I_{30}$ plotted in Figures 3–5. Whatever the cause of the cutoff, the points shown with central dots in Figure 4 by way of example, were rejected in computing the extrapolation lines shown. The rejection of such terminal "stray" points in the least-square fitting program can be implemented by the operator on the basis of his observations of maxima in the pulse profile (e.g., point X, Figure 3), and such points can be eliminated automatically within the computer program by statistical criteria, together with a small proportion of nonterminal stray points arising from a variety of disturbances.

4.2. Description of the Mark III PICS Instrument

A schematic diagram of the apparatus is given in Figure 7, and the two photographs (Figures 8 and 9) show, respectively, the front aspect of the PICS fluid bath in close-up and the same bath with the computer interface and analog recorder. The video display desk computer is a standard unit which is not shown here. Like the Mark II, the Mark III instrument is a "carousel" model, handling concurrently four sample cells which are mounted in a rotatable cross-shaped holder or carousel. This can be rotated through 90° very precisely by a stepping motor. One of the four cells will generally be in the fluid bath immersed in silicone oil, where a low-power He–Ne laser beam passes axially through it. The other three

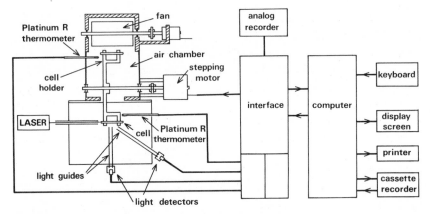

FIGURE 7. Schematic diagram of PICS apparatus. The rotatable cell holder or carousel carries four sample cells, two of which are shown, and one of which is aligned with the laser at any time, as shown.

cells rest in the air chamber above the fluid bath, where an even temperature well inside the stable range of the sample is maintained by means of an electric heater, fan, and Pt resistance thermometer. In Figure 8 the lid of the air chamber has been raised to expose the front arm of the carousel. The arm is seen via the access, used for mounting the cell in it at the beginning of a run.

FIGURE 8. Front aspect of PICS Bath showing (a) sliding door of Hot Box chamber, (b) fan drive, (c) stepping motor drive, (d) carousel arm with cell, (e) platinum resistance thermometer, (f) stirrer shaft, (g) observation window, and (h) laser.

FIGURE 9. PICS instrument showing (a) main bath assembly, (b) computer interface, (c) analog recorder.

The metal fluid-bath is heavily lagged against thermal disturbance from outside. It contains a stirrer, a heater, and cooling plates connected to an external circuit of cooling fluid. A Pt resistance thermometer measuring head is situated close to the sample cell in the fluid bath so as to measure its temperature to 0.01 °C. Two light guides collect the light scattered by the sample at angles centered at 30° and 90°, respectively, and transmit it to the photodiodes. The incident laser beam is also taken through a light guide close to the cell in order to reduce as much as possible the amount of stray incident light "seen" by the photodiodes, and to minimize dispersion of the incident beam by dust and bubbles in the bath fluid.

4.2.1. Computer-Controlled Flexibility

A computer-controlled four-channel multiplexer and analog-to-digital converter allows the programmed measurement of the light-scattering intensities, and of the cell and air-chamber temperatures. The stepping motor allows computer-controlled rotation of the carousel cell holder through very fast and highly reproducible single steps of $7° 30'$. The sequence control of intensity and temperature measurements, and the positioning of the cell holder in the laser beam, are determined entirely by programs written into the 32K random access memory of the (Commodore PET) computer of the PICS instrument. Since the temperature measurement is by Pt resistance, calibration programs which incorporate standard temperature–resistance relationships allow the calculation and continuous display of the temperature measurements. Operational and data analysis programs are chosen by a *menu control*. Considerable research flexibility is provided by the possibilities of extending or modifying existing programs.

4.2.2. Correction for Light Scattering by Density Fluctuations

A method used early in the development of PICS for determining the level of solvent or "base-line" scattering (due to density fluctuation) was to include a reference cell containing pure solvent as one of the four cells. Thus, in the data collected from a run, every fourth pulse height corresponded to the solvent scattering and this could, in principle, be deducted from total scattering of the polymer solutions in the other cells in order to correct for the intensity due to mere density (as distinct from concentration) fluctuations. However, because the cells were not optically matched, an empirical correction was found necessary.

A technique giving more reproducible results now exploits the fast measurement capability of the electronics. About 0.2 s after the cell holder has rotated to transfer a cell from the hot chamber to the cold bath and into the path of the laser beam and before the contents of the cell have cooled appreciably, the light intensities at both 30° and 90° detectors are measured. The intensity of the scattered light is thus determined before the solution cools enough to give rise to scattering by concentration fluctuations. This gives a value of each base-line intensity of light scattered at *high* temperature, even though the measurement is made while the cell is actually immersed in the cold bath!

After the base-line sampling, the solution quickly cools to the temperature of the bath, and after temperature equilibration (about 8–12 s) the new I_{30}, I_{90} and the temperature of the cell are measured. The timing of this second sampling is dependent on the programmed thermal pulse length chosen at the beginning of the run.

Thereafter the stepping motor rotates the cell holder through 90° at a time, and the process is repeated on successive cells in cyclic order. The computer subtracts the base-line measurement *for each cell* from its corresponding peak height and the critical scattering data at both angles for each cell are stored in an array together with their corresponding temperatures, while the chart recorder provides a continuous check (Figure 6).

4.2.3. Operational Sequence for a Run Determining Spinodal and Cloud-Point Temperatures

4.2.3a. Spinodals. The fluid bath is set to cool at $\sim0.05\,°C/min$ from a temperature a little above the highest cloud point expected for the four samples under investigation. The operator initiates the run. At his command, the computer automatically aligns cell number one to a position in the air chamber at 90° rotation from entry into the incident beam. As the temperature falls, all further measurements are made under programmed control. The operator will see the pulse heights increase (Figure 6) on the chart recorder, and from their shapes will determine when phase separation has begun in all four cells. He then terminates the run. The I_{30}, I_{90}, and T data for each cell are then printed out by the computer for examination prior to their analysis.

4.2.3b. Cloud Point. Before removing the cells, the cloud points can conveniently be measured. The bath is heated to near the temperature at which the sample of lowest T_s first showed phase separation through the I_{90} pulse shape (Y, Figure 6). That sample is then brought from the air chamber into the bath, and kept there while the bath temperature is set to rise at $\sim0.1\,°C$ per minute. The I_{90} value is sampled every few seconds and will be seen to go through a maximum (Figure 7 or Reference 1), at the cloud point, where the minute droplets dissolve practically instantaneously and scattering reverts to concentration fluctuation type. The procedure is repeated for the other three cells, in order of increasing cloud-point temperatures, as indicated by their pulse shapes.

5. CHARACTERIZATION OF POLYMERS BY SPINODALS IN SOLUTION: SOME GENERAL PRINCIPLES

Traditional and fundamental characterization procedures leave some gaps which measurements of critical points, spinodals, and cloud-point curves currently promise to fill.[4] Processing and some other technical properties respond extremely sensitively to higher moments of the MWD and to other structural features such as branching, which often cannot be quantitatively characterized by the fundamental procedures with sufficient accuracy to predict technical performance. Critical points, spinodals traced progressively outwards in both directions from their critical point, and finally cloud-point curves also pursued in increasing distances from the same point, provide a range of measurable thermodynamic quantities which stand in increasing order of sensitivity to higher moments[4] of the MWD, and to branching,[18,19,20] etc. What makes these nicely graded quantities of solution thermodynamics potentially interesting to technology is the way in which they are expected to reflect the technical behavior of bulk specimens. We may make this point clear by imagining that we have at our disposal a large number of polystyrene samples. To find two samples whose technological behavior in some process could not be distinguished, we would try to match their critical points, then the course of their spinodals outwards from that point, and finally, if necessary, even their cloud point curves over some sufficiently large neighborhood of the critical point also. If the matching process is carried sufficiently far, we should expect any bulk property, that may be of interest to the technologist, to be matched also.

Where several structural parameters exert sensitive influences, the ideal of their exact numerical characterization is not only difficult to attain, it can be counterproductive if carried to excess. In responding to practical needs, we are likely to do better by developing instrumental methods for characterization which reflect in their measurements the same kind of *coupling* between structural parameters as governs the practical properties concerned. The following discussion may help to make this clear.

Extensive theoretical considerations and experimental evidence suggest that the spinodals of *linear* polymers in solvents (at least in absence of substantial polarity or interactions) will be determined largely by M_w, M_z, and M_{z+1}, and possibly, but with rapidly decreasing effect, by the next few higher averages. The effects exerted by the averages on the location and shape of spinodals are coupled, so that a change in one of them can, to some extent, though not completely, be compensated by specific adjustment in the others. This kind of dilemma is familiar to those seeking methods of exact characterization. For instance, much fundamental work in polymer characterization has been intimately concerned with second virial coefficients B_2 measured at high dilution. However, it also became clear that there is generally strong statistical coupling between the coefficients in virial series, specifically also in polymer solutions, and recently Schäfer[21] was led to regard measurements of B_2 in such solutions to provide merely "effective" (rather than exact) values. The effective values incorporate the effects of higher virial coefficients[22] which are hard to eliminate with certainty.

For such reasons, the direct instrumental attack on the right combination of parameters, which tries to match the way they are coupled together in practice, may offer a better route to characterization. For instance, theories of such practical properties as die swell in extrusion involve higher moments of the MWD. The coupling effect between them, i.e., the way they combine to determine the physical behavior in bulk, may not be too far from the coupling between the same parameters in determining spinodals in solution, because they are essentially traceable to the same kind of interactions between segments operating under the same kind of chain configuration statistics. This heuristic claim rests on the circumstances that (a) near a critical point (and near a spinodal point close to the critical point) we have near-undisturbed chain configurations, as in the bulk, and (b) at such points the effects of the solvent molecules, and their interactions with the polymer segments and with each other, are largely compensated away—which is indeed the reason for the establishment of nearly undisturbed chain dimensions. The case of die-swell data in relation to PICS measurements of spinodals has been examined by Irvine.[4,23]

5.1. Characterization of Linear Polystyrene for Polydispersity with PICS

The best-known polymer/solvent system is polystyrene/cyclohexane (PS/CH). Because cloud-point curves are relatively easy to measure, the literature contains excellent data of this kind for PS/CH[24,56,26] and for many other systems. However, spinodals now being conveniently measurable, and much easier to interpret theoretically than cloud points, an extensive study on PS/CH spinodals was recently published.[3] The range of M_w covered was $5 \times 10^4 - 2 \times 10^6$ g mol^{-1}, and sharp fractions as well as highly heterodisperse mixtures were included; the range is extended below to about $M_w = 5 \times 10^6$ with some new data.

The primary purpose of the extensive study of spinodals lay in the need to refine further the free energy of mixing function $\Delta G = \Delta G(\phi, T)$ for the model system PS/CH, which typifies nonpolar polymer/solvent mixtures. The process of refinement was reviewed, back to its historic starting point with the Flory–Huggins model[8,9] in the early 1940s, and spinodal measurements were shown to have played an important role since they were used by Koningsveld et al.[27] in 1974 in confirming the validity of a new approach to reconciling theories for ΔG, applicable, respectively, in the dilute and semidilute concentration ranges. Bridging between the two ranges required an experimental technique in the right range of sensitivity to the parameters in ΔG, and spinodals were a good choice for this purpose. The refined bridging theory provided[3] a ΔG function (hereafter called ΔG_r, r = refined) whose implied critical points checked very well against published data from several sources, and whose implied cloud points fitted moderately well some measurements by Nakata et al.[25] close to the critical point. The refinement of the form of ΔG, a function which always fully characterizes the conceptual model since it implies the information on all measurable thermodynamic quantities, has currently been pushed to near the limit set by our ability to handle molecular interactions.[3] Adjustments in ΔG had decreased, in the course of refinements of the theory, down to amounts of only about $0.07RT$ for one whole polymer chain of 2 million in molecular weight, i.e., to about 1/20 of the kinetic energy of a gaseous atom at the same temperature!

The objectives of characterizing polymer samples must be sharply distinguished from that of refining a model theory by comparison with experiments. (For the general methodological distinction the reader may consult, e.g., Reference 28). Although both types of activity demand that the best ΔG function be used, the best may be different for the two different purposes. We now recall [Eq. (4)] the function developed and presented earlier by the refinement process, in which 275 spinodal measurements (ϕ_s, T_s) were employed in a least-squares fit to optimize five parameters $(\beta_0, \beta_1, \gamma, \lambda_0, \alpha)$, of which four were independently determinable from experiment or theory. The most satisfactory test for the soundness of the model refinement lay in the observed convergence of the values of these latter four parameters toward their independently known values under the least-squares optimization process, as the model was successively refined through the stages which retraced its history since the 1940s. The functional form of the up-to-date version of ΔG_r looks complicated:

$$
\begin{aligned}
\Delta G_r/RT = &\left(\sum_0^n \phi_i m_i \ln \phi_i \right) + [(\beta_0 + \beta_1/T)/(1 - \gamma\phi)]\phi_0\phi \\
&+ \phi_0 \sum_{i=1}^n \phi_i m_i^{-1} \exp\left[-\lambda \sum_{j=1}^N \phi_j m_j^{-1}(m_i^{1/2} + m_j^{1/2})^3 \right] \\
&\times [1 - \alpha \ln(m_i/\langle m_i \rangle_{\text{geom}})] \\
&+ [(1 - \gamma)\phi \ln(1 - \gamma) - (1 - \gamma\phi) \ln(1 - \gamma\phi)]\gamma^{-1} \quad (4)
\end{aligned}
$$

Here ϕ_i and m_i are volume fraction and reduced chain length (molecular weight/117.76) of the ith polymer molecule, $\phi = \sum_1^N \phi_i \equiv 1 - \phi_0$, and $\langle \ \rangle_{\text{geom}}$ means the geometric mean. The complexity of the form of ΔG_r is not serious, because in any case it must be handled by the computer. Computer programs are available to calculate from equations such as (4) the implied thermodynamic quantities, viz., spinodals, cloud-point curves, critical points, chemical potentials, etc. associated with phase equilibria. These programs are based on an approximation by specifying only the first r moments, say, of the desired MWD; but spinodals curves near ϕ_c converge rapidly with r increasing beyond 3 or 4. More precisely, the desired MWD is *replaced* by a notional mixture of delta function distributions (i.e., sharp fractions), viz., the unique mixture comprising the minimum number of sharp fractions which shares the first r moments with the desired MWD. This so-called *r-equivalent distribution*[2,29] is determined by an available computer program.[30] A two-component mixture of sharp fractions can thus be found which is 3-equivalent to the desired one, i.e., both share the first three moments and hence M_n, M_w, and M_z. This simplifies Eq. (4), in that the number of fractions n is reduced from virtually infinite to $N = 2$, and makes computations feasible.

5.2. The Currently Best ΔG for Characterizing Samples of Polystyrene (= ΔG$_{ch}$)

Two changes are made in the free energy of mixing function of Eq. (4), in light of the present knowledge on characterization. These changes are as follows:

(i) In the previous fittings of spinodals and of the Nakata cloud-point data,[3,31] the curves calculated from Eq. (4) with optimized parameters fell systematically below the experimental data for samples of high molecular weight. The correct reason for this was not recognized, because the theta temperature, i.e., the limit temperature for the spinodal at infinite MW, was then believed to be[24] 306.4 ± 0.2°K, while the value now quoted is appreciably higher,[32] viz., 308.6°K. The two parameters concerned with temperature behavior in Eq. (4), β_0 and β_1, are insufficient to match the data accurately over the wide temperature range of the available data. Accordingly we make the following substitution in Eq. (4) so as to introduce the additional purely empirical parameter β_2 (which is invoked in the thermodynamic literature as the next higher correction term; see Eq. (15) of Reference 33, where further references are given):

$$\beta_0 + (\beta_1/T) \to \beta_0 + (\beta_1/T) + \beta_2 T \qquad (5)$$

(ii) Additionally we set

$$\alpha \equiv 0 \qquad\qquad (6)$$

This is because introduction of β_2 was found to generate undesirable effects if α was retained. The optimized value of α was increased rather than decreased by introducing β_2, though the simplest theory has $\alpha = 0$. The presence of α tends to improve fittings at high dilution (as before), but it worsens fittings near the higher concentrations around the critical point, which are of special value for characterizing samples. In view of these findings, we prefer to eliminate α altogether [see Eq. (6)] by putting it equal to zero. Since in any case we thereby abandon the attempt to maintain optimum fitting for polydisperse samples data at high dilution, we also take the grave step of eliminating the three "worst-fitting" experimental points, which all represent very dilute solutions, out of the 275 previously used.[3] We would hesitate to take such a step if we were trying to *refine* a theoretical model. However, when *characterizing* unknown samples, better predictions based on measurements of spinodals close to the critical point are likely to result if we omit the worst points at high dilution, and if we replace the semiempirical parameter α in ΔG_r [Eq. (4)] by the purely empirical β_2. In Table 1 we compare the old parameters found in the refinement exercise with the new set proposed for characterizing unknowns. The new set was optimized over a data set of $275 - 3 = 272$ just mentioned plus 19 points belonging to two new polydisperse samples labeled 39 and 40 of higher M_w than occurred in the old set. These samples are described in Section 5.5 below. We first exemplify and discuss typical fittings of six experimental spinodals, including the two new ones, taken from the calibra-

TABLE 1. Optimized Parameters for ΔG_r (Eq. (4)) and for ΔG_{ch} Proposed for Characterization

	$\beta_0{}^a$	β_1	β_2	γ	λ_0	α	Standard deviation
ΔG_r	0.1717	100.53	0	0.4651	0.6215	0.025	8.6×10^{-4}
ΔG_{ch}	-1.85	398.1	0.00345	0.4585	0.5567	0	8.74×10^{-4}

[a] Note: The parameters β_0, β_1, γ_1, and λ_0 in ΔG_r were optimized over data belonging to 11 samples (sharp fractions) only, while α was subsequently optimized over the complete set of 38 samples. The parameters for ΔG_{ch} were optimized all together over the enlarged data set of 40 samples.

tion exercise of least-squares optimization of the new ΔG function (ΔG_{ch}), i.e., using the changes of Eqs. (5) and (6) in (4), over all 40 samples.

5.3. Sample Fittings from the Calibration of the Best Free-Energy Function, ΔG_{ch}, for Characterization of Polystyrene Samples

The six sample fittings in Figure 10 are labeled with their respective sample numbers as before.[3] Samples 14–16 and 34 present the data previously published, in comparison with optimized fittings to Eqs. (4)–(6). The critical points are not shown, but generally lie on the branch of the spinodal descending to the right of the maximum, somewhere near the middle of the range of concentrations $\varnothing$ covered by the experimental points. The fit of sample 34 is somewhat improved by changing from Eq. (4) to ΔG_{ch}. The differences between the original fittings (3) of samples 14–16 and the present ones in Figure 10 lie systematically in the directions indicated by the discussion of the old and new ΔG functions in Section 5.2. Thus the three calculated spinodals now fit *better* than before in the critical region along the right-hand descending branch while the fit in the dilute (left-hand) region has become *worse*. Thus slopes of the almost linearly descending

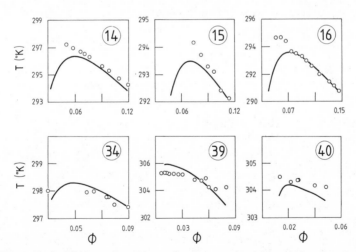

FIGURE 10. Typical fittings of experimental points to spinodal curves for six polydisperse samples of PS in CH, computed from ΔG_{ch} [Eqs. (4)–(6)], with parameters shown in Table 1. The independent data on the samples used are found in Table 2. The function ΔG_{ch} is designed to fit the data points best on the descending branch to the right of the maximum of each curve. Considering the scaling of the plots, which cover a temperature range of 13 degrees, the fittings of the descending branches are satisfactory; their steepness is a measure of polydispersity (cf. Table 2).

TABLE 2. Molecular Weight Averages of Samples Used in Figure 10

Sample number	$M_n \times 10^{-5}$ or Exper. $M_w/2$	$M_w \times 10^{-5}$	$M_z \times 10^{-5}$	$M_{z+1} \times 10^{-5}$
14	0.496	2.02	4.45	4.94
15	0.39	1.14	3.77	4.86
16	0.41	1.15	3.18	4.01
34	2.01	2.44	3.02	3.53
39	24.0	48.0	72.0	96.0
40	8.35	16.7	25.0	33.4

branches are seen in Figure 10 to be excellently predicted by the theoretical curves. Reference to the original fittings reveals that there all three theoretical curves descended rather more steeply than the data points. By way of contrast, the regrettable marked systematic downturn at the sensitive low concentration end of the theoretical curves, when compared with the data points in plots 14, 15, 16, is not observed in the original fittings to Eq. (4). There the fit was much improved, in the dilute-solution range, by the minute change in the form of ΔG_r induced by the value (0.025, Table 1) of α.

The power for characterization is well documented in the examples of Figure 10. They cover (Table 2) the 50-fold range of M_w from 5×10^6 down to 10^5 (but the fit is still good at 5×10^4). The polydispersity expressed as M_{z+1}/M_n, ranges from 1.8 to 12.5. We choose this unconventional measure of wide-span polydispersity as appropriate to our purpose. While M_n has practically no influence on spinodals in our range, and was disregarded in testing the model Eq. (4) in favor of M_{z+1} which exerts a greater influence, somewhat better predictions of the spinodals of unknown samples are found if the 3-equivalent mixtures are calculated on the basis of the measured averages M_n, M_w, M_z, (rather than M_w, M_z, and M_{z+1}). This is no doubt due to the fact that M_n is much more reliably measured, and thus provides a better input datum, than M_{z+1}. The samples in Figure 10 were either carefully weighed mixtures of commercially available sharp fractions (numbers 14, 15, 16, 34) or of especially well-characterized Flory distributions (numbers 39 and 40; see Section 5.5). The former class provides the best access to samples of reliable higher MW averages (calculated from those of the sharp-fraction components and their weight fractions in the mixture): the second class, that of Flory (most-probable or geometric) distributions, is now accessible as the first directly synthesizable quasicontinuous MWD. This more realistically simulates that of technological products, and its higher molecular weight averages are characterizable with accuracy comparable to that obtainable for sharp fractions. This is because any one average determines all the others if the MWD is known to be of Flory type, a precondition which can be checked by chemical kinetics and by the measurement of three different averages of the MWD.

A further check is provided by light-scattering, e.g., by the linearity, specific to the Flory distribution, of the scattering envelope in the Zimm plot based on traditional light-scattering experiments. This linearity is also beneficial to the accuracy of the extrapolation to determine M_w from the intercept of the Zimm plot.

The fitting of the Flory-distributed samples 39 and 40 in Figure 10 is thought to be very acceptable. True, there is a systematic displacement of the spinodal curve calculated from ΔG_{ch} in sample 40, and both 39 and 40 reveal a tendency for the calculated descending branch of the spinodal to descend somewhat too steeply; by comparison, no such displacement or error in slope is discernible in the descending branch of the spinodal of sample 16. Errors in placement and slope, which predominantly reflect errors in M_w and in the higher averages, respectively, occur in a number of our 40 samples (e.g., 14 and 15) besides the two Flory-distributed samples 39 and 40. In our judgment, these errors are too small in relation to the total ranges covered to offer hopes for improvement by further refinement of the theoretical ΔG function. On the contrary, we believe such errors to be traceable for the most part to errors in our independent estimates of the actual MWDs of the samples whose spinodals were being measured, and which enter the calculation of the theoretical spinodal curves through the estimated $M_n - M_z$. The 3-equivalent mixture of the two sharp (delta function) fractions, whose theoretical spinodals are actually plotted in Figure 10, match the estimates of these three averages for the sample under test, but may differ (usually mildly) in M_{z+2}, though this can hardly be ascertained at present by direct measurements. These considerations lead to the strong conjecture that the experimental spinodal points, such as presented in Figure 10, would correlate better with important processing variables than calculations based on the best available estimates of the average molecular weights. This is the more likely since both spinodals and technical variables like die swell are dependent to some extent on higher averages which are too difficult to measure directly, and since theory suggests (see Section 5) that the dependence may involve similar mathematical combinations of these averages.

5.4. Practical Exploitation of Thermodynamic Measurements on Polystyrene/Cyclohexane, Using ΔG_{ch}

Although the proposed free energy of mixing function ΔG_{ch} [Eq. (4)] as amended in Eqs. (5) and (6), provides a *fundamental* tool for characterizing PS samples, implementation in *practical* situations can be much simplified by preparing diagrams containing families of calibration plots for the required ranges of molecular weight averages. Computer programs based on ΔG_{ch} can be supplied for this purpose. Other formats, such as nomographs or superposition laws (master curves), could be developed in the future.

5.5. The Flory-Distributed Samples 39 and 40 (Figure 10)

These two samples were kindly made available by Professor G. Meyerhoff (Mainz). They represent thermally polymerized PS thoroughly investigated by Appelt and Meyerhoff.[34] Kinetic studies and polydispersity measurements have shown convincingly that these samples obey the most probable or Flory type of MWD. In particular, the polydispersity measurement $U \equiv (M_w/M_n) - 1$ may be calculated for their data to be 0.967 (theory 1.0) with standard deviation 0.218, using the 17 measurements given for their eight samples by their three different methods (Table IV of Reference 34). The concept of characterization in terms of moments of the MWD, directly averaged by physical measurement, has been exploited skillfully at Mainz. In particular, Hack and Meyerhoff[35] remark that light-scattering techniques can be as effective as GPC in determining U which illustrates the way in which averaging by direct physical measurement complements the wholesale fractionation by GPC, followed by computer averaging over the MWD spectrum (where base-line problems become serious when higher moments are to be determined). The work at Colchester reported here and earlier extends the philosophy of Meyerhoff to averages beyond M_w.

Indeed, by extending the spinodal measurements on sample 39, to concentrations in cyclohexane well below 1%, we provide data (Figure 11) which depend on higher moments too sensitively to be useful at present for characterization. The same remark applies, with even more force, to the cloud-point measurements also shown in Figure 11. However, even at present such measurements can serve for the qualitative comparison of samples, or equivalently for monitoring degradation of a given sample as we explain below. First, a description of measurements on sample 39 in

Figure 11, whose $M_w = 4.8 \times 10^6$ g mol^{-1} is the highest for which PICS measurements have been published, will be useful. Figure 3 shows that typical PICS Debye–Scholte plots (cf. Figures 4, 5) are of satisfactory quality to locate T_s by extrapolation in dilute solutions of a polymer of MW $\sim 5 \times 10^6$. The experimental spinodal points plotted in Figure 10, sample 39, were a subset chosen by objective methods from those in Figure 11. To have included the *full set* would have defeated our purpose of optimizing the parameters in fitting ΔG_{ch} to the data set of 40 samples, since the other samples comprised much smaller numbers of points (cf. Figure 10 and Figure 1 of Reference 3). A statistical reweighting of the data set according to the number of points comprised by various samples was thought too cumbersome in comparison with the adopted selection of points from sample 39.

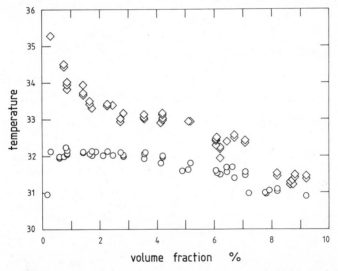

FIGURE 11. Spinodal ($\bigcirc$) and cloud ($\diamondsuit$) points for sample 39 (Table 2, cf. Figure 10) in cyclohexane. Only points near the highest temperatures observed for a given volume fraction are shown, because degradation effects can lower T_s and T_p. The upper edges of both sets of points define smooth curves. The critical point is believed to be in the range 6%–10%. The T_s value at $\sim 31\ °C$ of the solution of lowest concentration has a possible error of about $\pm 1/2°$. Since the spinodal curve is necessarily tangent to the temperature axis at absolute zero temperature, the spinodal curve must be nearly vertical near this point (on the scale here plotted), so that the error is of no practical consequence. Note that the even steeper descent of the cloud point curve as volume fraction $\phi \to 0$ has escaped detection. The steep *increase* in cloud-point temperature T_p with decreasing concentration, in the lowest concentration range accessible to measurement, is an exceedingly sensitive consequence of polydispersity.[3] The highest T_p shown may be above the theta point!

The spinodal points presented in Figure 11 are, in fact, themselves a selection from a much larger set of measurements actually made. A considerable scatter of results, unexplained at first, was observed. On the assumption that the scatter was due to degradation of the sample, which would lead systematically to reduction in M_w and hence in T_s, the points in Figure 11 were selected from the highest T_s observed at a given concentration. The explanation of the scatter (some of which remains in Figure 11 even after the elimination of the lower T_s values) in terms of degradation effects seemed plausible, because mechanical degradation of samples of MW larger than about 5×10^6 during the preparation of solutions is a known hazard. It was confirmed that storage of a given PICS cell over a period of weeks led to successively lower T_s measurements over a range of $\sim 0.5°$. The sample required periodic rehomogenization (administered in a centrifugal homogenizer[36]), but it is not known yet whether the degradation effect was indeed purely mechanical or involved oxidative or other kinds of scission, singly or in combination. We hope to investigate this further, since the monitoring of the spinodal curve provides a sensitive means of studying the degradation mechanism.

6. CONCLUSIONS

Although for convenience we have described the phase diagrams (Figure 1) and typical measurements for the case of upper critical solution temperature (UCST) behavior, PICS can equally be applied to LCST systems (Figure 12). Although both classes show qualitatively similar features apart from the obvious inversion of maxima to minima, the relevant refinements of the free energy function are known to require very different molecular effects to be considered. Thus a ΔG function of the form of Eq. (4) is intended for UCST systems of low polarity only. The process of characterizing samples can, of course, be applied at a sufficiently empirical level without reference to free energy functions altogether, but in this review we have been concerned to emphasize progress from the more fundamental perspective.

The instrument under review was shown to make contact with fundamental thermodynamic theory (see also the Appendix). From a more practical viewpoint, much of the instrument's power lies in the choice from continuous ranges of sensitivity which it offers to the experimenter, as he pursues spinodals and then cloud points from their common source at the critical point in both directions, toward lower and higher polymer concen-

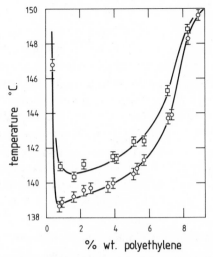

FIGURE 12. PICS measurements of spinodals (□) and cloud points (○) for a system with lower critical solution point behavior. Linear polyethylene of $M_w = 177$ kg/mol in *n*-hexane was measured under the vapor pressure of the solutions of ~ 5 atm.

trations. The almost inexhaustible reserve in sensitivity has been exploited here to *display* errors, rather than to *mask* them by contracting the scales of the plots.

The technique has been outlined in a number of textbooks,[37,38] and applications have been given in a variety of contexts.[40–53] The rationale behind the interplay between theory and experiment in the combined process of refining thermodynamic models (free energy of mixing functions) and of characterizing the necessary samples has been discussed in a recent review.[4] The process is here illustrated by the passage from ΔG_r ($r =$ refinement) to ΔG_{ch} (ch = characterization). Polydispersity has previously been singled out in a number of reviews,[18,39,3] but chain branching has been shown to be extremely sensitively reflected in spinodal curves,[18,19,20] a circumstance which can be exploited for characterization. Other microstructural features are also bound to be very noticeably reflected in PICS measurements, so that much work remains to be done.

APPENDIX: CONVEXITY AND STABILITY

A1. Two-Component Systems

Consider plots of free energy against concentration of solute (Figure 2). If diffusion leads locally to a separation of two regions, one more concentrated and one more dilute than the original homogeneous solution, the free energy will move vertically from the *curve* to the *chord* spanning the two concentrations. The driving force for restoring homogeneity accordingly depends on positive curvature (convexity downwards) of the plot; for then the free energy of the sum of the two regions will be *higher* than that for the original homogeneous solution, because the chord spanning the two concentrations lies *above* the original convex-downward curve. At a spinodal point, where the curve is locally straight, the chord and the curve coincide, so that there is no net rise in free energy when the two regions are formed, and the fluctuation must be expected to grow very large. In the unstable concentration domain (dotted segments in Figure 2), any fluctuation leads to decrease in free energy, so no barrier exists against phase separation which is then called spinodal decomposition.

Before proceeding from two-component to multicomponent systems, we collect some further thermodynamic remarks about Figure 2. The general shape of the three plots is dependent on the molecular weights of the two components only in a relatively trivial way; the greater the discrepancy in their molecular weights, the greater the dissymmetry of the diagram, in the sense that for a solution of high-molecular-weight polymer in an ordinary solvent, the critical and spinodal points (C, S) shown would lie very close to the axis $\phi = 0$. The bottom curve is everywhere *convex* downwards and this forms a condition of *stability* of the single-phase solution. A further thermodynamic constraint requires that all such plots approach a vertical course at both ends, i.e., at $\phi = 0$ and $\phi = 1$. This steep rise in G_{mol} simply reflects the resistance which nature opposes to purification, viz., to the removal of the last traces of either polymer or solvent. The point on the bottom curve, denoted (ϕ_c, T_c), is a critical point where the curvature of the convex curve takes the value zero, the lowest value compatible with convexity. Curves plotted for progressively lower temperatures must have progressively higher G_{mol}, since following the laws of thermodynamics, $(\partial G_{\mathrm{mol}}/\partial T)_p = -S_{\mathrm{mol}} < 0$. We may expect to find the more sinuous curves already discussed, with regions of negative and regions of positive curvature, such as those marked T_2 and T_3, lying *above* the curve with the critical point at T_c.[†]

The sinuosities "growing out" of the critical point recall similar effects in P/V diagrams of gas/liquid systems near their critical points. Quite

---

[†] We are taking the case of a lower critical solution temperature (LCST) as throughout. For a critical point related to an upper critical solution temperature (UCST), the sinuous curves would lie *below* the bottom curve.

generally, statistical mechanical models furnish theoretical plots of free energy against concentration (or P against V, etc.), which must then be examined by the theoretician for regions of convexity and concavity, if the stability as a single phase of the system modeled is to be understood.

The classical *double-tangent* construction restores the convexity of the two top curves by adding the solid line shown to short-circuit the broken (metastable) and dotted (unstable) sections. This construction furnishes the molar free energy plot for *stable* two-phase systems in the concentration range of the linear section, tangent at its two end points to the original sinuous curve. The two end points, i.e., the points of tangency, mark the composition of the two binary phases in equilibrium at the *cloud point* for the given temperature.

A2. Multicomponent Systems, e.g., Solutions of Polydisperse Polymers

In mixtures of more than two components, the simple measure $d^2G/d\phi^2$ for the curvature is no longer a measure of the reciprocal scattering intensity, but it has to be generalized to the quantity appropriate for measuring curvature in a multidimensional composition space. In that space, a complete set of independent volume fractions ϕ_i ($i = 1, 2, \ldots, n$) of n components (usually homodisperse polymer fractions) act as coordinates, leaving the $(n + 1)$th component (usually the solvent, denoted by 0) determined through the normalization

$$\phi_0 + \sum \phi_i = 1$$

The free energy G may be thought of as a "surface" (by adding the G coordinate in the direction of the nth dimension), and the important quantity measuring curvature and scattering intensity is the Hessian determinant, here termed Gibbs's spinodal determinant $|\partial^2G/\partial\phi_i\partial\phi_j|$. This vanishes precisely wherever the free energy surface becomes locally "flat" by lacking curvature in some *one* direction of composition space. (The reader may picture a saddle point or pass on a three-dimensional mountain. If the free energy surface in a three-component system has such a shape, the mixture is unstable with respect to separation of some possible pairs of phases, but stable to other separations, depending on the directions at the saddle point in which the mountain surface is convex or concave downwards, respectively).

Accordingly, at a spinodal point, Gibbs's equation,

$$
| \partial^2 G / \partial \phi_i \partial \phi_j | \equiv
\begin{vmatrix}
\dfrac{\partial^2 G}{\partial \phi_1{}^2} \ , & \dfrac{\partial^2 G}{\partial \phi_1 \partial \phi_2} \ , & \cdots \ , & \dfrac{\partial^2 G}{\partial \phi_1 \partial \phi_n} \\[2ex]
\dfrac{\partial^2 G}{\partial \phi_2 \partial \phi_1} \ , & \dfrac{\partial^2 G}{\partial \phi_2{}^2} \ , & \cdots \ , & \dfrac{\partial^2 G}{\partial \phi_2 \partial \phi_n} \\[1ex]
& \cdots & & \\[1ex]
\dfrac{\partial^2 G}{\partial \phi_n \partial \phi_1} \ , & \dfrac{\partial^2 G}{\partial \phi_n \partial \phi_2} \ , & \cdots \ , & \dfrac{\partial^2 G}{\partial \phi_n{}^2}
\end{vmatrix}
\qquad \text{(A1)}
$$

is satisfied, and such points form a continuous level surface in composition space, the *spinodal locus*. Since Gibbs's time, linear algebra has been strongly developed, and forms the best approach to computations of spinodal and critical behavior (see References 2 and 30).[†]

REFERENCES

1. K. W. Derham, J. Goldsbrough, and M. Gordon, *Pure Appl. Chem.* **38**, 97–116 (1974).
2. M. Gordon, P. Irvine, and J. W. Kennedy, *J. Polym. Sci. Polym. Symp.* **61**, 199–220 (1977).
3. M. Gordon and P. Irvine, *Macromolecules* **13**, 761–772 (1980).
4. H. Galina, M. Gordon, P. Irvine, and L. A. Kleintjens, IUPAC Symposium, 13–17 July (1981), *Pure Appl. Chem.* **54**, 365 (1982).
5. P. Debye, *J. Chem. Phys.* **31**, 680 (1959).
6. Th. G. Scholte, *J. Polym. Sci. A2* **8**, 841 (1970).
7. Th. G. Scholte, *J. Polym. Sci. C* **39**, 281 (1972).
8. P. J. Flory, *J. Chem. Phys.* **31**, 680 (1959).
9. M. L. Huggins, *Ann. N. Y. Acad. Sci.* **43**, 1 (1942).
10. J. W. Gibbs, *Trans. Conn. Acad.* **3**, 108, 343 (1876/8); reprint in *The Scientific Papers of J. W. Gibbs*, Vol. I, Dover, New York (1961).
11. P. Debye, H. Coll, and D. Woermann, *J. Chem. Phys.* **33**, 1746 (1960).
12. P. Debye, B. Chu, and D. Woermann, *J. Chem. Phys.* **36**, 1803 (1962).
13. B. Chu, *J. Chem. Phys.* **47**, 3816 (1976); B. Chu, F. J. Schoenes, and M. E. Fisher, *Phys. Rev.* **185**, 219 (1969).
14. W. Burchard and K. Kajiwara, *Proc. R. Soc. (London) Ser. A* **316**, 185–199 (1970).
15. J. M. G. Cowie, J. Goldsbrough, M. Gordon, and B. W. Ready, BP 1, 377, 478: DP 2, 161, 555; USP 3, 807, 865.
16. J. W. Kennedy, M. Gordon, and G. A. Alvarez, *Polimery*, Warsaw, No. 10, 464–471 (1975).

[†] For a recent review with information on, and references to, the use of PICS in the study of polymer compatibility, see R. Koningsveld and L. A. Kleintjens, in *Polymer Blends and Mixtures*, P. J. Walsh, J. S. Higgins, and A. Maconnachie, eds., Martinus Nijhoff Publishers, Dordrecht, Boston, Lancaster, 1985, pp. 89–116; see also M. Gordon, *ibid.*, p. 429.

17. R. Koningsveld and J. A. Staverman, *J. Polym. Sci. A2* **6**, 349 (1968).
18. K. W. Derham and M. Gordon, *Proc. of Polymerphysik* (Ch. Ruscher, ed.), Ges. der DDR, Berlin (Symposium held in Leipzig, 11–14 September 1974).
19. L. A. Kleintjens, R. Koningsveld, and M. Gordon, *Macromolecules*, **13**, 303 (1980).
20. L. A. Kleintjens, Ph. D. thesis, Essex (1979).
21. L. Schäfer, private communication (to be published).
22. K. R. Roberts, J. A. Torkington, M. Gordon, and S. B. Ross-Murphy, *J. Polym. Sci. Polym. Symp.* **61**, 45–62 (1977).
23. P. Irvine, Ph. D. thesis, University of Essex (1979).
24. N. Kuwahara, M. Nakata, and M. Kaneko, *Polymer* **14**, 415–479 (1973).
25. N. Nakata, T. Dobashi, N. Kuwahara, M. Kaneko, and B. Chu, *Phys. Rev. A.* **18**, 2683 (1978).
26. J. Hashizume, A. Teramoto, and H. Fujita, *J. Polym. Sci. Polym. Phys. Ed.* **19**, 1405–1422 (1981).
27. R. Koningsveld, W. H. Stockmayer, J. W. Kennedy, and L. A. Kleintjens, *Macromolecules* **7**, 73 (1974).
28. J. W. Kennedy, M. Gordon, J. Essam, and P. Whittle, *J. Chem. Soc. Faraday Trans. 2* **73**, 1289–1307 (1977).
29. P. Irvine and M. Gordon, *Proc. R. Soc. London Ser. A* **375**, 397–408 (1981).
30. P. Irvine and J. W. Kennedy, *Macromolecules* **15**, 473–528 (1982).
31. M. Gordon and P. Irvine, *Polymer* **20**, 1450 (1979): Corrigenda, *Polymer* **21**, 472 (1980).
32. R. Slagowski, B. Tsai, and D. McIntrye, *Macromolecules* **9**, 687 (1976).
33. K. W. Derham, J. Goldsbrough, M. Gordon, R. Koningsveld and L. A. Kleintjens, *Makromol. Chem. Suppl.* **1**, 401 (1975).
34. B. Appelt and G. Meyerhoff, *Macromolecules* **13**, 657–662 (1980).
35. H. Hack and G. Meyerhoff, *Makromol. Chem.* **179**, 2475–2488 (1978).
36. J. A. Torkington, L. Kleintjens, M. Gordon, and B. W. Ready, *Br. Polym. J.* **10**, (1978); M. Gordon and B. W. Ready, USP 4, 131, 369.
37. O. Olabisi, L. M. Robeson, and M. T. Shaw, *Polymer–Polymer Miscibility*, Academic Press, New York (1979).
38. G. Glöckner, *Polymercharakterisierung durch Flüssigkeitschromatographie*, VEB Deutscher Verlag der Wissenschaften, Berlin (1980).
39. M. Gordon, J. Goldsbrough, B. W. Ready, and K. Derham, *Industrial Polymers: Characterized by Molecular Weight* (J. H. S. Green and R. Dietz, eds.), pp. 45–51, Transcripta Books, National Physical Laboratory, London (1973).
40. R. Koningsveld and L. A. Kleintjens, *Pure Appl. Chem. Macromol. Chem. Suppl.* **8**, 197 (1973).
41. R. Koningsveld, L. A. Kleintjens, and H. M. Schoffeleers, *Pure Appl. Chem.* **39**, 1 (1974).
42. R. Koningsveld, *Br. Polym. J.* **7**, 435 (1975).
43. L. A. Kleintjens, H. M. Schoffeleers, and L. Domingo, *Br. Polym. J.* **8**, 29 (1976).
44. R. Koningsveld and L. A. Kleintjens, *J. Polym. Sci. Polym. Symp.* **61**, 221 (1977).
45. R. Koningsveld and L. A. Kleintjens, *Br. Polym. J.* **9**, 213 (1977).
46. R. Koningsveld, *Ber. Bunsenges.* **81**, 960 (1977).
47. L. A. Kleintjens, H. M. Schoffeleers, and R. Koningsveld, *Ber. Bunsenges.* **81**, 980 (1977).
48. R. Koningsveld and R. F. Stepto, *Macromolecules* **10**, 1166 (1977).

49. R. Koningsveld, *Bull. Soc. Chim. Beograd.* **44**, 5 (1979).
50. M. H. Onclin, L. A. Kleintjens, and R. Koningsveld, *Makromol. Chem. Suppl.* **3**, 197 (1979).
51. R. Koningsveld, L. A. Kleintjens, and M. H. Onclin, *J. Macromol. Sci. Phys. B* **18** (3), 363 (1980).
52. L. A. Kleintjens, M. H. Onclin, and R. Koningsveld, *EFCE Publ. Ser.* **11**, Proc. Berlin Conf., Part II, 521 (1980).
53. R. Koningsveld, M. H. Onclin, and L. A. Kleintjens, *Proc. MMI*, MMI Press Symp. Ser., 2 (Polym. Compat. Incompat.), 25–58 (1982).

Index